BOTANICAL RESEARCH AND PRACTICES

NEW INSIGHTS ON PLANT SEX CHROMOSOMES

BOTANICAL RESEARCH AND PRACTICES

Additional books in this series can be found on Nova's website
under the Series tab.

Additional E-books in this series can be found on Nova's website
under the E-book tab.

BOTANICAL RESEARCH AND PRACTICES

NEW INSIGHTS ON PLANT SEX CHROMOSOMES

RAFAEL NAVAJAS-PÉREZ

EDITOR

Nova Biomedical Books
New York

Library of Congress Cataloging-in-Publication Data

New insights on plant sex chromosomes / editor: Rafael Navajas-Pirez.
    p. cm.
  Includes index.
  ISBN 978-1-61470-236-8 (softcover)
  1. Plant chromosomes. 2. Sex chromosomes. I. Navajas-Pirez, Rafael.
  QK725.N49 2011
  571.2--dc23
                2011022213

Published by Nova Science Publishers, Inc. †New York

Contents

vi

Preface

One of the most important topics in evolutionary biology concerns the origin and evolution of sex-determining systems and sex chromosomes. Certain plant species display younger sex-chromosome systems in different evolutionary stages. It is thought that the same evolutionary forces described for animals are operating in plants. However, in opposition to the situation in animals, sex-determining mechanisms seem to be more flexible, and most species with separate sexes have evolved directly from ancestors with both sex functions. These features make plants excellent models for studies on sex determination. In this context, early evolved plant-sex chromosomes have given rise to many studies in recent years. In this book, the most recent findings are highlighted and reviewed, focusing specifically on model species, including Carica papaya, Fragaria virginiana, Silene latifolia and Rumex acetosa.

Chapter 1 - Sex chromosomes are thought to have evolved from an autosomal pair through the accumulation of sex-determining genes and the disruption of X-Y recombination that ultimately led to the formation of heteromorphic sex chromosomes. Early evolved plant-sex chromosomes have given rise to many studies in recent years, which have proved chromosomal rearrangements and repetitive-DNA accumulation crucial events that play an important role in sex-chromosome evolution. In this review, the authors combine evidence gathered by different approaches and report the most recent findings in three model species on plant sex-chromosome analyses, including different levels of sex-chromosome differentiation: *Carica papaya*, homomorphic sex chromosomes; *Silene latifolia*, heteromorphic and undifferentiated; and *Rumex acetosa*, heteromorphic and highly differentiated.

The data are discussed in the light of plant sex-chromosome differentiation and evolution.

Chapter 2 - Sex chromosomes evolved from a pair of autosomes. During their evolution, suppression of recombination occurred in the sex determining regions, the nonrecombining regions then expanded resulting in the production of the heteromorphic sex chromosomes. In humans and *Drosophila melanogaster*, the sex chromosomes developed about 300 million years ago (mya) and 60 mya, respectively. By contrast, plant sex chromosomes are comparatively new; the papaya sex chromosomes developed 2–3 mya, and the sex chromosomes of the dioecious *Silene latifolia* arose 10–20 mya. The study of these young sex chromosomes allows us to understand the early steps of sex chromosome evolution. Repetitive sequences are thought to be involved in preventing recombination and in causing the expansion of the nonrecombining regions. Moreover, the location of repetitive sequences could provide clues into large-scale chromosomal rearrangements. In the last decade, many repetitive sequences have been characterized in *S. latifolia* and they have been used as probes for fluorescent *in situ* hybridization. These studies showed that large rearrangements were involved in sex chromosome evolution in *S. latifolia*, and that Y chromosome-specific repetitive sequences accumulated during its evolution. Together with the data from the genetic mapping of sex chromosome-linked genes, these results are very informative for understanding the structure of the sex chromosomes and the process of their evolution.

Chapter 3 - According to the currently accepted model, sex chromosomes evolve from a pair of autosomes via a series of step-wise expansions of the non-recombining region on the Y chromosome (NRY), creating 'evolutionary strata', followed by degeneration of genes trapped in the NRY. This model is based on the analysis of ancient (over 200 million year [MY] old) sex chromosomes of mammals and birds, as well as recent translocations of autosomes to sex chromosomes (neo-sex chromosomes) in *Drosophila*. Much less is known about plant sex chromosomes, with the bulk of data coming from only two species, *Silene latifolia* and *Carica papaya*, and even for these best studied plants, less than a dozen sex-linked genes have been identified and analysed. Here the author critically discusses the main conclusions from the data available. In particular, he stresses that the evidence for 'evolutionary strata' and for genetic degeneration of Y-linked genes in plants is relatively weak. Thus, it is premature to extend the 'animal model' of sex chromosome evolution to plants, as the evolutionary trajectory of plant sex chromosomes may differ from that in animals.

Sex chromosomes are peculiar structures present in most animal and some plant genomes [1]. Although they evolved many times independently, their properties are quite similar in different groups of organisms. The chromosome restricted to one sex (Y in male- or W in female-heterogametic species) lacks recombination and is genetically degenerate, while its counterpart (X or Z, respectively) is present in both sexes, actively recombines in the homogametic sex and is relatively gene-rich. Independent evolution of similar structures in different organisms provides a wealth of material for evolutionary genetic studies seeking to understand the origin and evolutionary forces involved in sex chromosome evolution.

Chapter 4 - Papaya is a semi-woody tree that produces fruit rich in vitamins and minerals. It is trioecious with male, female, and hermaphrodite plants. Though many theories have been suggested in the past, papaya sex is determined by a pair of nascent sex chromosomes; females have two X chromosomes, males have an X and a Y, and hermaphrodites have an X and a Y^h, which varies slightly from the male Y. Any combination of the Y and Y^h genotype, YY, Y Y^h, or Y^h Y^h, is lethal. The X and Y chromosomes have a small non-recombining region in the centromeric and pericentromeric region. The hermaphrodite-specific region of the Y^h chromosome (HSY) is gene poor and has an increased amount of retroelements and chromosomal rearrangements compared to its X counterpart and the genome wide average. The X and Y chromosomes were estimated to have diverged relatively recently about 2-3 million years ago (MYA), and even more so for the Y and Y^h (73,000 years ago). Physical maps of the HSY and the corresponding X region have been produced and sequenced, showing the HSY sequence has expanded. Genes in these regions are being mined with a special focus on identifying the two sex determination genes, one promoting maleness and one suppressing femaleness. The impact of identifying the sex determination genes would be high for both the commercial production of papaya and the field of sex chromosome evolution.

Chapter 5 - *Fragaria* is an exceptional model system for understanding sexual system and sex chromosome evolution. The genus hosts the entire range of sexual systems, as well as novel features, such as diversity of ploidy levels and the presence of female heterogametey. Comparative genetic mapping has revealed the autosomal ancestor of sex chromosomes in two species and exposed the evolutionary lability of sex-determining chromosomes in this young system. Recent QTL analyses have allowed the first glimpse of the genomic architecture of secondary sexual dimorphism and point the way for future work aimed at understanding the driving forces behind sex chromosome initiation and turnover. The genus is also poised to aid in answering the question of how genome doubling and merger facilitates the initiation of polymorphic sexual systems and the development of sex chromosomes.

Chapter 6 - The origin and evolution of sexual dimorphism and the sex-determining mechanisms are major topics in Evolutionary Biology on which many studies have focused in recent decades. Among flowering plants, the origin of dioecy has resulted from quite recent events, occurring independently in about 7% of the genera. However, only a moderate number of dioecious plant species exhibit chromosome-mediated sex-determination systems. The genus *Rumex* (*Polygonaceae*), with monoecious, gynodioecious, hermaphroditic, and polygamous representatives, and with dioecious species bearing sex chromosomes in different evolutionary stages (homomorphic, XX/XY, XX/XY$_1$Y$_2$), has considerably contributed to shed light on this topic. Although still patchy, current knowledge on sex-chromosome evolution has greatly benefited from analyses on species of this group, for which the most significant findings are reviewed in this chapter.

In: New Insights on Plant Sex Chromosomes   ISBN: 978-1-61470-236-8
Editor: Rafael Navajas-Pérez   ©2012 Nova Science Publishers, Inc.

Chapter I

Plant Sex-Chromosome Evolution

Pedro J. Sola-Campoy, Carmelo Ruiz Rejón,
Roberto de la Herrán, and Rafael Navajas-Pérez[*]
Departamento de Genética, Facultad de Ciencias,
Universidad de Granada, Granada, Spain

ABSTRACT

Sex chromosomes are thought to have evolved from an autosomal pair through the accumulation of sex-determining genes and the disruption of X-Y recombination that ultimately led to the formation of heteromorphic sex chromosomes. Early evolved plant-sex chromosomes have given rise to many studies in recent years, which have proved chromosomal rearrangements and repetitive-DNA accumulation crucial events that play an important role in sex-chromosome evolution. In this review, we combine evidence gathered by different approaches and report the most recent findings in three model species on plant sex-chromosome analyses, including different levels of sex-chromosome differentiation: *Carica papaya*, homomorphic sex chromosomes; *Silene latifolia*, heteromorphic and undifferentiated; and *Rumex acetosa*, heteromorphic and highly differentiated. The data are discussed in the light of plant sex-chromosome differentiation and evolution.

[*] Corresponding author: Departamento de Genética, Facultad de Ciencias, Universidad de Granada, Campus de Fuentenueva s/n, 18071. Granada, SPAIN, email: rnavajas@ugr.es

INTRODUCTION

Reproduction can be as simple as a cell dividing (asexual reproduction) or can involve the combination of genetic material from two taxonomically related organisms through sexual reproduction. In terms of energy cost, sexual reproduction can jeopardize proliferation, although it is preponderant all through the fungus, plant, and animal kingdoms. The importance of sexual reproduction lies in ensuring a new genetic combination by avoiding self fertilization, and decreasing deleterious mutation rates by recombination, reasons advantageous enough to develop strong mechanisms for sex-determination maintenance, such as sex chromosomes bearing sex-determining genes [1, 2].

Table 1. Glossary of most common terms related to sex determination in plants.

Glossary

Hermaphrodite	plants with both male and female reproductive parts in the same flower
Dioecious	plants in which female and male reproductive organs are separated in different individuals
Androdioecious	plant species in which male and hermaphrodite flowers are borne on separate individuals
Gynodioecious	plants that bear female and hermaphrodite flowers on separate individuals
Monoecious	plants in which female and male reproductive organs are separated in different floral structures on the same plant
Andromonoecious	plant species in which male and hermaphrodite flowers are borne separately on the same individual
Gynomonoecious	plants that bear female and hermaphrodite flowers on the same individual
Trioecious	plants that bear unisexual (dioecious) and hermaphrodite flowers in different individuals
Polygamous	general term to describe plants that bear hermaphrodite and unisexual flowers on the same individual or different

Sex determination in plants is controlled by genetic factors and, in some species, can be influenced by growth hormones and environmental factors. Contrary to the situation in animals, such mechanisms seem to be more flexible, and related representatives with intermediate sexual conditions are found (see **Table 1**). Dioecy is a rare condition in plants (38% of all angiosperms and very few gymnosperms, bryophita and algae), and only around 40 representatives of such species have sex chromosomes (**Table 2** [3, 4].

Table 2. List of most representative plant species with sex chromosomes

Family	Species	Sex chromosome	Sex det. mechanism	References
		Angioperms		
Actinidiaceae	*Actinidia deliciosa*	Heterogametic male	Active-Y	[74]
	Actinidia chinensis	Heterogametic male	Active-Y	[75]
Amaranthaceae	*Acnida sp.*	Heterogametic male	Active-Y	[3]
Asparagaceae	*Asparagus officinalis*	Heterogametic male	Active-Y	[76]
Asteraceae	*Antennaria dioica*	Heterogametic male	-	[]3
Cannabidaceae	*Cannabis sativa*	♀XX/ ♂XY	X/A ratio	[77]
	Humulus lupulus	♀XX/ ♂XY	X/A ratio	[78]
	Humulus japonicus	♀XX/ ♂XY$_1$Y$_2$	X/A ratio	[79]
Cariaceae	*Carica papaya*	Heterogametic male	Active-Y	[80]
	Vasconcellea sp.	Heterogametic male	Active-Y	[80]
Caryophyllaceae	*Silene latifolia*	♀XX/ ♂XY	Active-Y	[81]
	Silene dioica	♀XX/ ♂XY	Active-Y	[15]
Chenopodiaceae	*Spinacia oleracea*	Heterogametic male	Active-Y	[82]
Cucurbitaceae	*Bryonia dioica*	Heterogametic male	Active-Y	[66]
	Ecballium elaterium	Heterogametic male	Active-Y	[3]
	Coccinia indica	♀XX/ ♂XY	Active-Y	[83]
Dioscoreaceae	*Dioscorea tokoro*	Heterogametic male	Active-Y	[84]
Euphorbiaceae	*Mercurialis annua*	Heterogametic male	Active-Y	[3]
Polygonaceae	*Rumex acetosa* group	♀XX/ ♂XY$_1$Y$_2$	X/A ratio	[85, 52]
	Rumex acetosella	♀XX/ ♂XY	Active-Y	[49]
	Rumex hastatulus	♀XX/ ♂XY or ♂XY$_1$Y$_2$	Active-Y- X/A ratio	[49]
	Rumex suffruticous	♀XX/ ♂XY	Active-Y	[17]
Ranunculaceae	*Thalictrum sp.*	Heterogametic male	-	[3]
Rosaceae	*Fragaria vesca*	Heterogametic female	Active-W	[86]
	Fragaria chiloensis	Heterogametic male	Active-Y	[86]
	Fragaria virginiana	Heterogametic male	Active-Y	[86]
Salicaceae	*Populus nigra*	♂ZZ/♀ZW	Active-W	[67]
Vitiaceae	*Vitis sp.*	♀XX/ ♂XY	Active-Y	[63]
		Gymnosperms		
Cycadaceae	*Cycas revoluta*	♀XX/ ♂XY	Active-Y	[87]
	Cycas pectinata	♀XX/ ♂XY	Active-Y	[69]
Ginkgoaceae	*Ginkgo biloba*	♀XX/ ♂XY	Active-Y	[68]
Podocarpaceae	*Podocarpus macrophyllus*	♀XX/ ♂XXY	-	[88]
Zamiaceae	*Zamia sp.*	Heterogametic male	-	[89]
		Bryophyta		
Marchantiaceae	*Marchantia polymorpha*	♀XX/ ♂XY	Active-Y	[71]
Sphaerocarpaceae	*Sphaerocarpos donnellii*	♀XX/ ♂XY	-	[90]
		Algae		
Phaeophyceae	Ectocarpus siliculosus	-	-	[72]

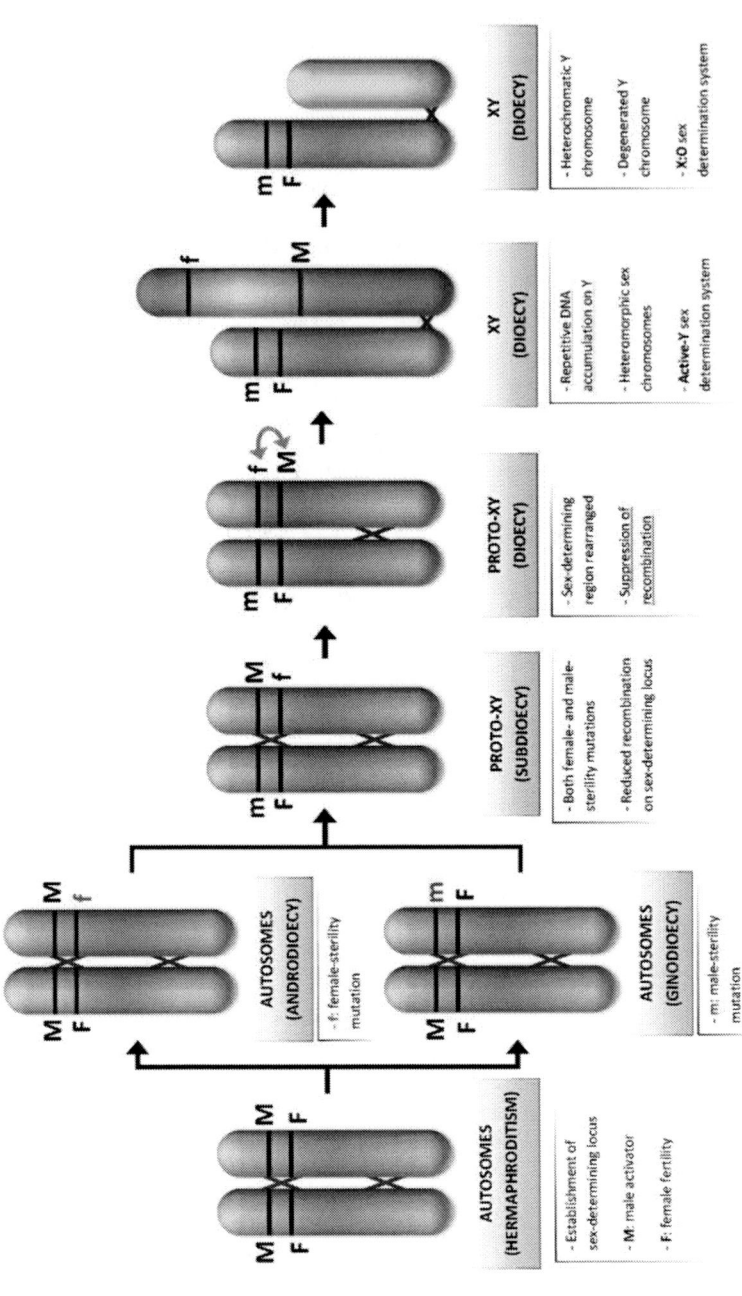

Figure 1. Evolutionary pathway to dioecy from an hermaphroditic ancestor through androdioecious or gynodioecious intermediates, and establishment of sex-determining region and pseudo-autosomal recombining regions, orange stands for suppression of recombination and yellow represents repetitive and transposable DNA element accumulation in Y chromosome.

In dioecious representatives, sex chromosomes are thought to have evolved from a standard autosomal chromosome pair as a consequence of a rarely recombining region containing genes involved in sex determination, followed by Y-chromosome degeneration [5]. Dioecy has evolved separately in the evolution by a stepwise process that implies three evolutionary stages [5]. In some cases this derives from monoecious ancestors through andromonoecious or gynomonoecious intermediates, by the genetic fixation of sex ratios between individuals [6] , and more frequently, from hermaphroditic ancestors through gynodioecious or androdioecious intermediates, the latter being extremely rare in plants (**Figure 1** [7, 8]). The first stage involves the establishment of the sex-determining locus (or loci) on an autosomal pair. At least two mutations, male-sterility and female-sterility mutations, are required (**Figure 1**). In the second stage, the suppression of recombination on the sex-determining region, which is favoured mainly by some sort of chromosomal rearrangement, triggers the molecular degeneration of Y chromosomes (or W in WZ systems). In the final third stage, recombination occurs only in a small region because X and Y chromosomes are too diverged, the Y chromosome remains highly degenerated and heterochromatic (see **Figure 1**).

That progressive suppression of recombination promoted by chromosomal rearrangements (i.e. local duplications, inversions, and translocations; [9, 10]), is the ultimate consequence of the accumulation of diverse repetitive sequences, such as mobile elements and satellite DNAs. This may further inhibit recombination between X and Y chromosomes and ensure the maintenance of dimorphic sex chromosomes, while conferring them with exceptional evolutionary features. This suppression of recombination has been crucial for the rise and evolution of sex chromosomes and has occurred separately in different plant lineages [11].

The theoretic importance of the sex-chromosome study lies in the understanding of sex-determining mechanisms, this information also having practical application in crop improvement. Plant sex chromosomes are particularly informative because they have evolved much more recently than those of mammals or *Drosophila* (**Figure 2**), allowing access to early stages in the process of sex-chromosome emergence. Another fact particular to plants is that most species with separate sexes have evolved directly from ancestors with both sex functions. In this context, the enormous increase in studies on plant sex determination during the last decade is not surprising (see for example reviews by Ming and Moore [12] or Charlesworth [13]). The most intensively studied plants are *Carica papaya* [14], *Silene latifolia* [15], and *Rumex acetosa* [16, 17]. In this chapter, we review the most important findings regarding those models.

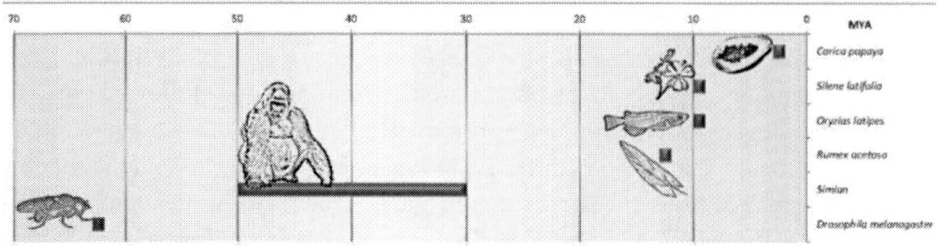

Figure 2. Scaled-temporal diagram representing the origin of sex-chromosomes system in model species based on molecular data: insects, *Drosophila melanogaster* [91]; fishes, *Oryzias latipes* [92]; mammalians, simians [45]; and flowering plants, *Rumex acetosa* (XX/XY1Y2 system) [16], *Silene latifolia*[15] , and *Carica papaya* [22].

DIFFERENT STAGES ON SEX-CHROMOSOMES EVOLUTION

1. Papaya (*Carica Papaya*), Brand-New Sex Chromosomes

The molecular mechanism of sex determination in papaya is not fully known, but the papaya male-specific Y region (MSY) may contain a female-suppressing gene and a male-fertility gene. Some authors have suggested that sex determination in this species is determined by a single gene corresponding to SEX1 locus, which has three alleles: two dominant alleles for males (SEX1-M) and hermaphrodites (SEX1-H) over the third one for females (*sex1-f*) [18, 19]. This locus is involved in stamen and carpel development/suppression and also it is hypothesised to bear genes responsible for cross-over suppression (C) and lethality (L), both present in the dominant alleles, giving rise to a 2:1 segregation for sex type [18]. Sex determination in papaya is mediated by the presence of a pair of sex chromosomes [14]. Thus, this species is polygamous with 2n=16 + XX in females, and XY in males or XY^h in hermaphrodite individuals. Papaya sex chromosomes are morphologically indistinguishable and represent an early stage on sex-chromosome evolution.

The centromere of the papaya Y chromosome has mapped within the MSY region [20]. This centromeric domain has accumulated significantly more DNA than the corresponding X chromosomal locus, which has contributed significantly to a partial cease of recombination, leading to abnormal chromosome pairing. Fine genetic mapping analyses on the MSY region and its counterpart in the X chromosome have shown a reduced number of expected genes and an accumulation of retroelements in the Y chromosome [21, 22]. Also, comparisons of BACs [23] and AFLP-segregation data have demonstrated that the rate of change in this region is significantly higher than in other papaya genomic regions (reviewed in [24]). It is noteworthy that homozygotes for the MSY region (YY individuals) are lethal, suggesting that some essential gene is lost, but, according to the current data, this might be due to lethal mutations in one gene or a few genes, or to silencing by methylation, rather than to a massive molecular degeneration. Methylation is thought to trigger the suppression of recombination in some sex chromosomes [25]. In fact, although some signs of heterochromatinization are visible in the MSY region in form of five knob-like heterochromatic blocks, the meiotic behaviour of sexual pairs is not affected except for the small region around the centromere [20]. The sex-determining region in papaya is still small, representing only 13% of Y chromosome, and no massive chromosomal rearrangements that could have resulted in a complete suppression of recombination are detected [20]. Other well-studied plants, such as asparagus (*Asparagus officinalis*, [26]), for which YY genotypes have been described, kiwi (*Actinidia deliciosa*, [27]) or wild strawberry (*Fragaria virginiana*, [28]) might show a similar situation; indistinguishable homomorphic sex-chromosomes, signs of suppression of recombination around a sex-determining region, and no clear evidence of molecular degeneration of Y chromosome.

A first draft of the whole genome sequence has been recently published for papaya [29]. This will benefit comparative analyses and will enable the characterization of sex-determining gene(s) when the Y chromosome is fully sequenced. To date, comparative analyses have been made using a BAC-by-BAC comparison bearing sequences from MSY region and its X counterpart. This has allowed estimates of the age of papaya sex chromosomes of about 0.5-2.5 million years ago (mya) (**Figure 2**), while the divergence between Y and Y^h is estimated to be 73000 years [23]. Similar analyses using the sister genus *Vasconcellea* dated the origin of papaya sex chromosomes at approximately 2-3 mya [30].

2. White Campion (*Silene Latifolia*), on the Way to Y-Chromosome Degeneration

Silene latifolia has 2n=22 autosomes + XY in males and XX in females. Sex chromosomes in white campion are morphologically distinguishable (heteromorphic) and constitute the largest pair of the karyotype; the Y chromosome being about 50% longer than the X chromosome and containing approximately the 9% of the diploid genome [31]. *S. latifolia* sex-determining system has been extensively studied and is possibly the best-known plant model in this area to date. At least three different groups of genes have been established in the Y chromosome: groups A and D, located in the short arm and related to gynoecium suppression and stamen-promoting activity, respectively; and group BC, located in long arm and related to stamen development and male fertility [32]. About a dozen Y-chromosome-linked genes and their counterparts in the X chromosome have been isolated such SlY1, SlY3, SlY4, DD44Y, and SlssY (Revised in [33] and [34]). However, genes expressed preferentially or exclusively located in Y chromosome have not been characterized so far. Only SlAP3Y (a MADS box gene) has been demonstrated to be involved in male-specific flower-bud maturation, but this gene has counterparts present in the X chromosome and in one autosome [35].

There is no recombination along almost the entire length of X and Y chromosomes except for a short region, the pseudoautosomal region (PAR), located on the distal end of the X short arm and the Y long arm [36]. Although the non-recombining process might have started shortly before, a large pericentric inversion triggered the gradual process of recombination restriction and led to Y-chromosome erosion [15, 33]. Some evidence points in this direction; thus, Marais et al. (2008) [37] showed a significantly faster evolution in four Y-linked genes than did their counterparts in X-linked. Another sign of degeneration comes from the fact that some introns of Y-linked genes are longer than those copies in the X chromosome. A detailed examination of the larger intron shows sequences belonging to transposable elements such LRT retrotransposons present in DD44Y or inverted repeats in SlY3 intron [37]. Transposable elements, together with tandem repeats, microsatellites, and chloroplast DNA have in fact accumulated in the no-recombining region, playing an important role in sex-chromosome differentiation [38, 39]. The *S. latifolia* YY genotype is not viable, further indicating deterioration of genes not related to maleness, as is predicted for the sex-chromosome-evolution model [40].

However, recent data would support the contention that the degeneration of the Y chromosome is still in an early stage. Concretely, a hermaphrodite-inducing mutation has been proved to be located in the gynoecium-suppression region of the Y chromosome of *S. latifolia*, suggesting a lack of strong degeneration of the Y chromosome [41]. Also, Qiu et al. (2011) [42], analysing the patterns of codon-usage bias in X- and Y-linked genes in *Silene latifolia*, found similar results.

Based on comparisons between X- and Y-linked genes, and considering a mutation rate of 10^{-8} mutations per base pair per generation, sex-chromosomes of *S. latifolia* have been dated 5-10 mya [43], this resembling findings in other plant species (**Figure 2**).

3. Sorrel (*Rumex Acetosa*), a Y-Chromosome Dead End

The last stage in the process of sex-chromosome formation comprises not only the suppression of recombination but also the accumulation of a repertory of repetitive elements -mainly transposable elements and satellite DNAs- in the Y chromosome. This final step gives rise to a dead-end molecular differentiation between X and Y chromosomes accompanied by dramatic morphological differences, i.e. heteromorphism. As a result, the Y chromosome remains as a non-recombining desert that preserves exclusively genes that are crucial for the maleness [44]. Most known sex-chromosome systems, including those of mammals and insects, reached this third stage after tens of millions of years of evolution (**Figure 2**; [45-47]). However, this terminal differentiation is highly unusual in organisms with younger sex-chromosome systems such as plants and fishes (revised in [48]).

In plants, *Rumex acetosa* and closely related species (the so-called *Acetosa* group) are characterized by the presence of an atypical XX/XY_1Y_2 sex-chromosome system that has been widely studied [49-54, 17]. This chromosomal complex, formed by unequal translocation or misdivision, is unique in the plant kingdom because both Y chromosomes are almost entirely heterochromatic, representing the most advanced stage in plant sex-chromosome evolution [11, 17]. In fact, the two Y chromosomes recombine only with their X-chromosome counterpart by a small PAR and do not recombine between them, constituting a sexual trivalent during meiosis. They are also almost fully heterochromatic and stain strongly with the fluorochrome DAPI [55]. A more detailed molecular survey reveals that Y-chromosome heterochromatin is constituted mainly by several satellite-DNA families, RAYSI, RAYSII, RAYSIII, and RAE180 [56-58] and several types of transposable elements [59]. Navajas-Pérez et al. (2009) [60] mapped comparatively RAYSI and RAE180 sequences by chromosome painting and characterized the structure of Y chromosomes in two related species, *R. acetosa* and *R. papillaris*. Thus, it was possible to infer that massive chromosomal rearrangements are taking place within Y-chromosome heterochromatin. This may have suppressed the recombination completely, leading to massive molecular degeneration. Although the accumulation of similar sorts of repetitive sequences has been described in sex chromosomes of other plant species such as in *Silene* or papaya [61, 21], the *R. acetosa* complex seems to have reached a deeper coverage and higher degeneration. In this context, considering the pattern of accumulation of such repetitive sequences, it is possible to infer different degrees of sex-chromosome differentiation in *Rumex;* species with undifferentiated homomorphic sex chromosomes, species with incipient XX/XY systems, and highly degenerated XX/XY$_1$Y$_2$ species (reviewed in [54]).

The *R. acetosa* sex-determination mechanism is *Drosophila*-like, depending on the ratio between X chromosomes and autosome sets [62]. By contrast, in papaya and white campion, sex-determination is human-like, controlled by the presence of a male-determining Y chromosome. The presence of *Rumex* relatives with both systems has offered proof that X/A ones evolved secondarily from male-determining Y ones [16], as suggested before by Westergaard (1958) [63]. This would represent a new piece of evidence that points to *R. acetosa* as the most evolved plant sex-chromosome system. Considering the mean rate of change in nuclear DNA of 0.6% per site per million years for plants [64], we used our data on internal spacer sequences of rDNA (ITS) to infer that dioecy would have appeared in *Rumex* between 15-16 mya, while the divergence time between the *R. acetosella/R. suffruticosus* (related group of species with XX/XY system) leading to the *Acetosa* XX/XY$_1$Y$_2$ clade should be 12-13 mya (**Figure 2**; [16]). Unfortunately, no candidate genes for sex-determining have been characterized so far and further analysis will be needed to keep clarifying the interesting sex-determining system of this group of species.

Finally, a last step in sex-chromosome evolution observed in animals is the dosage compensation of X-linked genes for maintaining correct gene-expression levels in females. That may have evolved as an adaptive response to the loss of gene activity on the Y chromosome. This stage has not been fully demonstrated, either in *Rumex* or in other plant species, although some level of differential methylation in one of the two X chromosomes in homogametic female cells has been reported in white campion [65].

CONCLUSIONS

Here, we have highlighted the most outstanding findings with regard to sex-chromosome systems in three model plant species: *Carica papaya*, papaya; *Silene latifolia*, white campion; and *Rumex acetosa*, sorrel. In all cases analysed, sex dimorphism and sex chromosomes dated between 2.5-13 mya (**Figure 2**). Although these estimates should be taken with caution due to the incompleteness of the fossil record and that different approaches and DNA sequences have been used to gather the data (reviewed in [73]), the results indicate that sex-chromosome systems in plants are very young. This, together with the fact that dioecious plants have evolved directly from ancestors with both sex forms, make plants excellent models to study sex determination and the evolution of the sex-chromosome systems.

Model plant species with sex-chromosomes parallel the three-step theory proposed by Charlesworth (1996) [5] for sex-chromosome evolution (**Figure 1**). An initial stage comprises species with undifferentiated homomorphic sex chromosomes. This stage, represented by papaya, asparagus or wild strawberry, is characterized by a small region involved in sex determination, incipient mechanisms for suppression of recombination such as methylation, and faint evidence of molecular erosion of Y chromosome chromatin, such as a low density of genes or accumulation of transposable elements in the MSY region. In a second stage, the extension of the region with suppressed recombination, normally promoted by drastic chromosomal rearrangements, becomes evident, leading to heteromorphic sex-chromosomes. However, the degree of Y molecular degeneration is still moderate. The maximum exponent to exemplify this situation is white campion. The third stage, represented by sorrel, includes highly differentiated sex chromosomes. In this stage, chromosomal rearrangements and massive accumulation of repetitive DNA are responsible of Y degeneration and the establishment of heteromorphic sex chromosomes.

As predicted theoretically, we have showed that in plants several mechanisms such as methylation, chromosomal rearrangements or the accumulation of repetitive elements may be actively contributing to sex-chromosome formation, triggering the suppression of recombination and favouring the subsequent molecular degeneration of Y chromosomes. It is noteworthy that, contrary to the situation in mammals, in all cases, there is a tendency for the Y chromatin to expand. This may further confirm that sex chromosomes in plants recently evolved and are still differentiating by heterochromatin expansion.

In short, sex chromosomes are found in different angiosperms, such as: *Carica papaya*, papaya; *Silene* spp. white campion; *Rumex* spp., sorrels; *Cannabis sativa*, hemp; *Actinidia deliciosa*, kiwi; or *Bryonia dioica*, red bryony [66]. In these species, the XX/XY system is preponderant but not exclusive; meanwhile, in *Populus nigra*, black poplar females are heterogametic ZW while males are homogametic ZZ [67]. Sex chromosomes have also been described in other groups such as: gymnosperms, *Ginkgo biloba*, ginkgo [68, 93] and *Cycas pectinata*, cycad [69]; mosses, *Ceratodon purpureus* [70]; hepatics, *Marchantia polymorpha* liverwort [71]; and algae *Ectocarpus siliculosus* [72] (**Table 2**). As the volume of genomic data from different plant species increases, the number of species with sex chromosomes is likely to increase as well. This opens a promising scenario for sex-determination studies, and will allow comparative analyses. The comparison of X- and Y-specific sequences will reveal chromosomal rearrangements and mutations of the sex-determining region after the suppression of recombination and will ultimately shed light on several aspects of plant sex chromosomes: i) to unravel chromatin rearrangements involved in X-Y differentiation, ii) to clarify the comparative evolution of sex-determining regions in diverse plant genomes, and ultimately iii) to characterize the genes responsible for sex determination in plants.

ACKNOWLEDGMENTS

This work was supported by project AGL2009-09094, P.J.S-C is a FPI scholar, both awarded by the Ministerio de Ciencia e Innovación, Spain.

REFERENCES

[1]   Goddard, M.R., Godfray, H.C., Burt, A. (2005). Sex increases the efficacy of natural selection in experimental yeast populations. *Nature, 434,* 636-640.

[2]   Paland, S., Lynch, M. (2006). Transitions to asexuality result in excess amino acid substitutions. *Science, 311,* 990-992.

[3]   Charlesworth, D., Guttman, D. (1999). The evolution of dioecy and plant sex chromosome systems. In: Ainsworth CC. ed. Sex Determination in Plants. Oxford. UK: BIOS Scientific Publishers, London: Society for Experimental Biology. pp. 25-49.

[4]   Ming, R., Bendahmane, A., Renner, S.S. (2011). Sex chromosomes in land plants. *Annual Review of Plant Biology, in press.*

[5]   Charlesworth, B. (1996). The evolution of chromosomal sex determination and dosage compensation. *Curr Biol, 6,* 149-162.

[6]   Renner, S.S., Ricklefs, R.E. (1995). Dioecy and its correlates in the flowering plants. *Am J Bot, 82,* 596-606.

[7]   Charlesworth, B., and Charlesworth, D. (1978). A model for the evolution of dioecy and gynodioecy. *The American Naturalist, 112,* 975-997.

[8]   Weiblen, G.D., Oyama, R.K., Donoghue, M.J. (2000). Phylogenetic Analysis of Dioecy in Monocotyledons. *Am Nat, 155,* 46-58.

[9]   Steinemann, M., Steinemann, S. (1997). The enigma of Y chromosome degeneration: TRAM, a novel retrotransposon is preferentially located on the Neo-Y chromosome of *Drosophila miranda. Genetics, 145,* 261-266.

[10]   Skaletsky H, Kuroda-Kawaguchi T, Minx P.J. et al. (2003). The male specific region of the human Y chromosome is a mosaic of discrete sequence classes. *Nature, 423,* 825-837.

[11]   Ruiz Rejón, M. (2004).Sex chromosomes in plants. In: Encyclopedia of Plant and Crop Sciences (Vol 6): Dekker Agropedia (6 vols), Marcel Dekker Inc., New York, pp. 1148-1151.

[12]   Ming, R., Moore, P.H. (2007). Genomics of sex chromosomes. *Curr Opin Plant Biol, 10,* 123-30.

[13]   Charlesworth D. (2008). Sex chromosome origins and evolution. In: Pagel M, Pamiankowski A, eds. Evolutionary genomics and proteomics. Sunderland (MA): Sinauer Associates, Inc. pp. 207-239.

[14]   Liu, Z., Moore, P.H., Ma, H., Ackerman, C.M., Ragiba, M., et al. (2004). A primitive Y chromosome in papaya marks incipient sex chromosome evolution. *Nature, 427,* 348-352.

[15]   Nicolas, M., Marais, G., Hykelova, V., Janousek, B., Laporte, V., Vyskot, B., Mouchiroud, D., Negrutiu, I., Charlesworth, D., Moneger, F. (2005).A gradual process of recombination restriction in the evolutionary history of the sex chromosomes in dioecious plants. *PLoS Biol, 3,* 47-56.

[16]   Navajas-Pérez, R., de la Herrán, R., López-González, G., Jamilena, M., Lozano, R., Ruiz-Rejón, C., Ruiz-Rejón, M., Garrido-Ramos, M.A. (2005). The evolution of reproductive systems and sex-determining mechanisms within *Rumex* (polygonaceae) inferred from nuclear and chloroplastidial sequence data. *Mol BiolEvol., 22,* 1929-1939.

[17]   Cuñado, N., Navajas-Pérez, R., de la Herrán, R., Ruiz Rejón, C., Ruiz Rejón, M., Santos, J.L., and Garrido-Ramos, M.A. (2007). The evolution of sex chromosomes in the genus *Rumex* (Polygonaceae): identification of a new species with heteromorphic sex chromosomes. *Chromosome Res, 15,* 825-832.

[18]   Storey, W.B. (1953). Genetics of papaya. *J Hered, 44,* 70-78.

[19]   Sondur, S.N., Manshardt, R.M., Stiles, J.I. (1996). A genetic linkage map of papaya based on randomly amplified polymorphic DNA markers. *Theor Appl Genet., 93,* 547-553.

[20]   Zhang, W., Wang X., Yu Q., Ming R., Jiang J. (2008). DNA methylation and heterochromatinization in the male-specific region of the primitive Y chromosome of papaya. *Genome Research, 18,* 1938-1943.

[21]   Yu, Q., Hou, S., Hobza, R., Feltus, F.A., Wang, X., et al. (2007). Chromosomal location and gene paucity of the male specific region on papaya Y chromosome. *Molecular Genetics and Genomics, 278,* 177-185.

[22]   Yu, Q., Hou, S., Feltus, F.A., Jones, M.R., Murray, J., Veatch, O., Lemke, C., Saw, J.H., Moore, R.C., Thimmapuram, J., Liu, L., Moore, P.H., Alam, M., Jiang, J., Paterson, A.H., Ming, R. (2008a). Low X/Y divergence in four pairs of papaya sex-liked genes. *Plant J., 53,* 124-132

[23]   Yu, Q., Navajas-Perez, R., Tong, E., Robertson, J., Moore, P.H., et al. (2008b). Recent Origin of Dioecious and Gynodioecious Y Chromosomes in Papaya. *Tropical Plant Biology, 1,* 49-57.

[24]   Charlesworth, D. (2004). Plant evolution: modern sex chromosomes. *CurrBiol, 14,* R271-R273.

[25]   Gorelick, R. (2005). Theory for why dioecious plants have equal length sex chromosomes. *Am J Bot, 92, 979-984.*

[26]   Telgmann-Rauber, A., Jamsari, A., Kinney, M.S., Pires, J.C., Jung, C. (2007). Genetic and physical maps around the sex-determining M-locus of the dioecious plant asparagus. *Mol Genet Genomics, 278,* 221-234.

[27]   Harvey, C. F., Fraser, L. G., and Gill, G. P. (1997). Sex determination in *Actinidia. Acta Hortic, 444,* 85-88.

[28]   Goldberg, M.T., Spigler, R.B., Ashman, T.L. (2010). Comparative genetic mapping points to different sex chromosomes in sibling species of wild strawberry (*Fragaria*). *Genetics, 186,* 1425-1433.

[29]   Ming, R., Hou, S., Feng, Y., Yu, Q., Dionne-Laporte, A., Saw, J.H., Senin, P., Wang, W., Ly, B.V., Lewis, K.L., Salzberg, S.L., Feng, L., Jones, M.R., Skelton, R.L., Murray, J.E., Chen, C., Qian, W., Shen, J., Du, P., Eustice, M., Tong, E., Tang, H., Lyons, E., Paull, R.E., Michael, T.P., Wall, K., Rice, D.W., Albert, H., Wang, M.L., Zhu, Y.J., Schatz, M., Nagarajan, N., Acob, R.A., Guan, P., Blas, A., Wai, C.M., Ackerman, C.M., Ren, Y., Liu, C., Wang, J., Wang, J., Na, J.K., Shakirov, E.V., Haas, B., Thimmapuram, J., Nelson, D., Wang, X., Bowers, J.E., Gschwend, A.R., Delcher, A.L., Singh, R., Suzuki, J.Y., Tripathi, S., Neupane, K., Wei, H.,

Irikura, B., Paidi, M., Jiang, N., Zhang, W., Presting, G., Windsor, A., Navajas-Pérez, R., Torres, M.J., Feltus, F.A., Porter, B., Li, Y., Burroughs, A.M., Luo, M.C., Liu, L., Christopher, D.A., Mount, S.M., Moore, P.H., Sugimura, T., Jiang, J., Schuler, M.A., Friedman, V., Mitchell-Olds, T., Shippen, D.E., dePamphilis, C.W., Palmer, J.D., Freeling, M., Paterson, A.H., Gonsalves, D., Wang, L., Alam, M. (2008). The draft genome of the transgenic tropical fruit tree papaya (*Carica papaya* Linnaeus). *Nature, 452,* 991-996.

[30]   Wu, X., Wang, J., Na, J.K., Yu, Q., Moore, R.C., Zee, F., Huber, S.C., Ming, R. (2010).The origin of the non-recombining region of sex chromosomes in *Carica* and *Vasconcellea*. *Plant J., 63,* 801-810.

[31]   Matsunaga, S., Hizume M., Kawano S., Kuroiwa T. (1994). Cytological analyses in *Melandrium album*: genome size, chromosome size and fluorescence *in situ* hybridization. *Cytologia, 59,* 135-141.

[32]   Zluvova, J., Janousek, B., Negrutiu, I., Vyskot, B. (2005). Comparison of the X and Y chromosome organization in *Silene latifolia*. *Genetics, 170,* 1431-1434.

[33]   Matsunaga, S. (2006).Sex chromosome-linked genes in plants. *Genes Genet. Syst., 81,* 219-226.

[34]   Howell, E.C., Armstrong, S.J., Filatov, D.A. (2011). Dynamic gene order on the *Silene latifolia* Y chromosome. *Chromosoma, in press.*

[35]   Matsunaga, S., Isono, E., Kejnovsky, E., Vyskot, B., Dolezel, J., Kawano, S., and Charlesworth, D. (2003). Duplicative transfer of a MADS box gene to a plant Y chromosome. *Mol. Biol. Evol, 20,* 1062-1069.

[36]   Lengerova, M., Moore, R. C., Grant, S. R., and Vyskot, B. (2003). The sex chromosomes of *Silene latifolia* revisited and revised. *Genetics, 165,* 935-938.

[37]   Marais, G.A., Nicolas, M., Bergero, R., Chambrier, P., Kejnovsky, E., Monéger, F., Hobza, R., Widmer, A., Charlesworth, D. (2008). Evidence for degeneration of the Y chromosome in the dioecious plant *Silene latifolia*. *Curr Biol, 18,* 545-549.

[38]   Hobza, R., Lengerova, M., Svoboda, J., Kubekova, H., Kejnovsky, E., Vyskot, B. (2006). An accumulation of tandem DNA repeats on the Y chromosome in *Silene latifolia* during early stages of sex chromosome evolution. *Chromosoma, 115,* 376-382.

[39]   Kejnovsky, E., Hobza, R., Cermak, T., Kubat, Z., Vyskot, B. (2009). The role of repetitive DNA in structure and evolution of sex chromosomes in plants. *Heredity, 102,* 533-541.

[40]   Vagera, J., Paulikova, D., Dolezel, J. (1994). The development of male and female regenerants by in-vitro androgenesis in dioecious plant *Melandrium album*. Ann Bot, *73,* 455-459.

[41]   Miller, P.M., Kesseli, R.V. (2011). A sex-chromosome mutation in *Silene latifolia*. *Sex Plant Reprod, in press*.

[42]   Qiu, S., Bergero, R., Zeng, K., Charlesworth, D. (2011). Patterns of codon usage bias in *Silene latifolia*. *MolBiolEvol.*, *28,* 771-780.

[43]   Bergero, R., Forrest, A., Kamau, E., Charlesworth, D. (2007). Evolutionary strata on the X chromosomes of the dioecious plant *Silene latifolia*: evidence from new sex-linked genes. *Genetics, 175,* 1945-1954.

[44]   Quintana-Murci, L., Fellous, M. (2001). The Human Y Chromosome: The Biological Role of a "Functional Wasteland". *J Biomed Biotechnol, 1,* 18-24.

[45]   Lahn, B.T. and Page, D.C. (1999). Four evolutionary strata on the human X chromosome. *Science, 286,* 964-967.

[46]   Ezaz, T., Stiglec, R., Veyrunes, F., Marshall Graves, J.A. (2006). Relationships between vertebrate ZW and XY sex chromosome systems. *Curr Biol, 16,* R736-R743.

[47]   Sánchez, L. (2008). Sex-determining mechanisms in insects. *Int J Dev Biol, 52,* 837-856.

[48]   Charlesworth, D., Charlesworth, B., Marais, G. (2005). Steps in the evolution of heteromorphic sex chromosomes. *Heredity, 95,* 118-128.

[49]   Löve, Á. (1957). Sex determination in *Rumex*. *Proc Genet SocCan, 2,* 31-36.

[50]   Smith B.W. (1969). Evolution of sex-determining mechanisms in *Rumex*. *Chromosomes Today, 2,* 172-182.

[51]   Degraeve, N. (1976). Contribution a létudecytotaxonomique des Rumex. IV Le genre *Acetosa* Mill. *La Cellule, 71,* 231-240.

[52]   Wilby, A.S., Parker, J.S. (1988). Recurrent patterns of chromosome variation in a species group. *Heredity, 61,* 55-62

[53]   Ainsworth, C.C., Lu, J., Winfield, M., Parker, J.S. (1999). Sex determination by X: autosome dosage: *Rumex acetosa* (sorrel). In: Ainsworth CC, ed. Sex Determination in Plants. Oxford: BIOS Scientific Publishers, pp. 124-136.

[54]   Navajas-Pérez, R. (2011). The genus *Rumex*: a plant model to study sex-chromosome evolution In: Navajas-Pérez, R. (ed), New Insights on Sex Plant Chromosomes, Nova Publishers, Hauppauge NY.

[55]   Ruiz Rejón, C., Jamilena, D., Garrido-Ramos, M.A., Parker, J.S., Ruiz Rejón, M. (1994). Cytogenetic and molecular analysis of the multiple sex chromosome system of *Rumex acetosa*. *Heredity*, *72*, 209-215.

[56]   Shibata, F., Hizume, M., and Kurori, Y. (1999). Chromosome painting of Y chromosomes and isolation of a Y chromosome-specific repetitive sequence in the dioecious plant *Rumex acetosa*. *Chromosoma*, *108*, 266-270.

[57]   Shibata, F., Hizume, M., and Kurori, Y. (2000). Differentiation and the polymorphic nature of the Y chromosomes revealed by repetitive sequences in the dioecious plant, *Rumex acetosa*. *Chromosome Res.*, *8*, 229-236.

[58]   Mariotti, B., Manzano, S., Kejnovský, E., Vyskot, B., and Jamilena, M. (2009). Accumulation of Y-specific satellite DNAs during the evolution of *Rumex acetosa* sex chromosomes. *Mol Genet Genomics*, *281*, 249-259.

[59]   Mariotti, B., Navajas-Pérez, R., Lozano, R., Parker, J.S., de la Herrán, R., Ruiz Rejón, C., Ruiz Rejón, M., Garrido-Ramos, M., and Jamilena, M. (2006). Cloning and characterization of dispersed repetitive DNA derived from microdissected sex chromosomes of *Rumex acetosa*. *Genome*, *49*, 114-121.

[60]   Navajas-Pérez, R., Schwarzacher, T., Ruiz Rejón, M., Garrido-Ramos, M.A. (2009). Molecular cytogenetic characterization of *Rumex papillaris*, a dioecious plant with an XX/XY_1Y_2 sex chromosome system. *Genetica*, *135*, 87-93.

[61]   Cermak, T., Kubat, Z., Hobza, R., Koblizkova, A., Widmer, A., Macas, J., Vyskot, B., Kejnovsky, E. (2008). Survey of repetitive sequences in *Silene latifolia* with respect to their distribution on sex chromosomes. *Chromosome Res*, *16*, 961-976.

[62]   Löve, A. (1969). Conservative sex chromosomes in *Acetosa*. In: Darlington C.D, Lewis K.R. Chromosomes Today, vol. 2. Plenum Press, New York. pp. 166-171.

[63]   Westergaard, M.(1958).The mechanism of sex determination in dioecious flowering plants. *Adv. Genet*, *9*, 217-281.

[64]   Gaut, B.S. (1998). Molecular clocks and nucleotide substitution rates in higher plants. *Evol.Biol*, *30*, 93-120.

[65]   Vyskot, B., Siroky, J., Hladilova, R., Belyaev, N.D., Turner, B.M. (1999). Euchromatic domains in plant chromosomes as revealed by H4 histone acetylation and early DNA replication. *Genome*, *42*, 343-350.

[66]   Oyama, R.K., Volz, S.M., Renner, S.S. (2009). A sex-linked SCAR marker in *Bryonia dioica* (Cucurbitaceae), a dioecious species with XY sex-determination and homomorphic sex chromosomes. *J Evol Biol.*, *22*, 214-224.

[67]   Yin, T., Difazio, S.P., Gunter, L.E., Zhang, X., Sewell, M.M., Woolbright, S.A., Allan, G.J., Kelleher, C.T., Douglas, C.J., Wang, M., Tuskan, G.A. (2008). Genome structure and emerging evidence of an incipient sex chromosome in Populus. *Genome Res*, *18*, 422-430.

[68]   Lee, C. L. (1954). Sex chromosomes in *Ginkgo biloba*. *Am. J. Bot*, *41*, 545-549.

[69]   Abraham, A., and Mathew, P.M. (1962). Cytological studies in the Cycads: Sex chromosome in *Cycas*. *Ann. Bot*, *26*, 261-267.

[70]   McDaniel, S.F., Willis, J.H., Shaw, A.J. (2007). A linkage map reveals a complex basis for segregation distortion in an interpopulation cross in the moss *Ceratodon purpureus*. *Genetics*, *176*, 2489-2500.

[71]   Okada, S., Fujisawa, M., Sone, T., Nakayama, S., Nishiyama, R., Takenaka, M., Yamaoka, S., Sakaida, M., Kono, K., Takahama, M., Yamato, K. T., Fukuzawa, H., Brennicke, A. and Ohyama, K. (2000). Construction of male and female PAC genomic libraries suitable for identification of Y-chromosome-specific clones from the liverwort, *Marchantia polymorpha*. *The Plant Journal*, *24*, 421-428.

[72]   Cock, J.M., Sterck, L., Rouzé, P., Scornet, D., Allen, A.E., Amoutzias, G., Anthouard, V., Artiguenave, F., Aury, J.M., Badger, J.H., Beszteri. B., Billiau, K., Bonnet, E., Bothwell, J.H., Bowler, C., Boyen, C., Brownlee, C., Carrano, C.J., Charrier, B., Cho, G.Y., Coelho, S.M., Collén, J., Corre, E., Da Silva, C., Delage, L., Delaroque, N., Dittami, S.M., Doulbeau, S., Elias, M., Farnham, G., Gachon, C.M., Gschloessl, B., Heesch, S., Jabbari, K., Jubin, C., Kawai, H., Kimura, K., Kloareg, B., Küpper, F.C., Lang, D., Le Bail, A., Leblanc, C., Lerouge, P., Lohr, M., Lopez, P.J., Martens, C., Maumus, F., Michel, G., Miranda-Saavedra, D., Morales, J., Moreau, H., Motomura, T., Nagasato, C., Napoli, C.A., Nelson, D.R., Nyvall-Collén, P., Peters, A.F., Pommier, C., Potin, P., Poulain, J., Quesneville, H., Read, B., Rensing, S.A., Ritter, A., Rousvoal, S., Samanta, M., Samson, G., Schroeder, D.C., Ségurens, B., Strittmatter, M., Tonon, T., Tregear, J.W., Valentin, K., von Dassow, P., Yamagishi, T., Van de Peer, Y., Wincker, P. (2010). The *Ectocarpus* genome and the independent evolution of multicellularity in brown algae. *Nature, 465,* 617-621.

[73]   Filatov, D. (2011). How much do we know about evolution of sex chromosomes in plants? In: Navajas-Pérez, R. (ed), New Insights on Sex Plant Chromosomes, Nova Publishers, Hauppauge NY.

[74] Ferguson, A.R., Seal, A.G., McNeilage, M.A., Fraser, L.G., Harvey, C.F., Beatson, R.A. (1996). Kiwifruit. In: Janick J, Moore JN eds. Fruit breeding. vol 2. Vine and small Fruits crops. J Wiley & Sons, New York, pp 371-417.

[75] Testolin, R., Cipriani, G., Costa, G. (1995). Sex segregation ratio and gender expression in the genus *Actinidia*. *Sex Plant Reprod.*, *8,* 129-132.

[76] Löptien, H. (1979). Identification of the sex chromosome pair in asparagus (*Asparagus officinalis* L.). *Z Pflanzenzüchtg*, *82*, 162-173.

[77] Yamada, I. (1943). The sex chromosome of *Cannabis sativa* L. *SeikenZiho*. *2*, 64:68.

[78] Ono, T. (1937). On sex-chromosomes in wild hops. *The Bot. Magazine*, *51*, 110-115.

[79] Kihara, H. (1929). The sex chromosomes of *Humulus japonicus*. *IdengakuZasshi*, *4*, 55-63.

[80] Horovitz, S., and Jimenez, H. (1967). Cruzamientos interespecificos e intergenericos en Caricaceas y sus implicaciones fitotecnicas. *Agronomia Tropical*, *17*, 323-344.

[81] Ono, T. (1939). *Polyploidy and sex determination in Melandrium. The Bot. Magazine, 53, 550-556.*

[82] Bemis, W.P., and Wilson, G.B. (1953). A new hypothesis explaining the genetics of sex determination in *Spinacea oleracea* L. *J Hered*, *44*, 91-95.

[83] Kumar, L. S., Vishveshwaraiah, S. (1952). Sex mechanism in *Coccinia indica* Wight and Arn. *Nature, 170,* 330-331.

[84] Martin, F.W., and Ortiz, S. (1963). Chromosome numbers and behavior in some species of *Discorea*. *Cytologia*, *28*, 96-101.

[85] Ono T. (1935). Chromosomen und Sexualität von *Rumex acetosa*. *Sci Rep Tohoku Imp Univ IV, 10,* 41-210.

[86] Kihara, H. (1930). Karyologische Studien an Fragaria mit besonderer Berücksichtigung der Geschlechtschromosomen. *Cytologia, 1,* 345-357.

[87] Segawa, M., Kishi, S., Tatuno, S. (1971). Sex chromosomes of *Cycas revoluta*. *Japan J. Genet.*, *46,* 33-39.

[88] Hizume, M., Shiraishi, H., Tanaka, A. (1988). A cytological study of *Podocarpus macrophyllus* with special reference to sex chromosomes. *Japan J. Genet, 63,* 413-423.

[89] Sangduen, N., Toahsakul, M., Hongtrakul, V. (2007). Karyomorphological Study of Some Selected Cycads. *AU J.T., 11,* 1-6.

[90] Allen, C.E. (1917). A chromosome difference correlated with sex in *Sphaerocarpos*. *Science, 46,* 466-467.

[91]   Carvalho A.B., Koerich, L.B., Clark, A.G. (2009). Origin and evolution of Y chromosomes: *Drosophila* tales. *Trends Genet*, *25*, 270-277.

[92]   Kondo M, Nanda I, Hornung U, Schmid M, Schartl M. (2004). Evolutionary origin of the medaka Y chromosome. *CurrBiol*, *14*, 1664-1669.

[93]   Pollock, E.G. (1957). The sex chromosome of the maiden hair tree. *J.Heredity*, *48*, 290-293.

In: New Insights on Plant Sex Chromosomes   ISBN: 978-1-61470-236-8
Editor: Rafael Navajas-Pérez   ©2012 Nova Science Publishers, Inc.

Chapter II

The Role of Repetitive Sequences in the Evolution of Plant Sex Chromosomes

Yusuke Kazama[*1] *and Sachihiro Matsunaga*[2]

[1]Radiation Biology Team, RIKEN Nishina Center, Saitama, Japan
[2]Department of Applied Biological Science,
Faculty of Science and Technology,
Tokyo University of Science, Chiba, Japan

ABSTRACT

Sex chromosomes evolved from a pair of autosomes. During their evolution, suppression of recombination occurred in the sex determining regions, the nonrecombining regions then expanded resulting in the production of the heteromorphic sex chromosomes. In humans and *Drosophila melanogaster*, the sex chromosomes developed about 300 million years ago (mya) and 60 mya, respectively. By contrast, plant sex chromosomes are comparatively new; the papaya sex chromosomes developed 2–3 mya, and the sex chromosomes of the dioecious *Silene latifolia* arose 10–20 mya. The study of these young sex chromosomes allows us to understand the early steps of sex chromosome evolution. Repetitive

[*] Corresponding author. Yusuke Kazama, Radiation Biology Team, RIKEN Nishina Center, 2-1 Hirosawa, Wako, Saitama 351-0198, Japan, Tel: +81-; Fax: +81-; E-mail: ykaze@riken.jp

sequences are thought to be involved in preventing recombination and in causing the expansion of the nonrecombining regions. Moreover, the location of repetitive sequences could provide clues into large-scale chromosomal rearrangements. In the last decade, many repetitive sequences have been characterized in *S. latifolia* and they have been used as probes for fluorescent *in situ* hybridization. These studies showed that large rearrangements were involved in sex chromosome evolution in *S. latifolia*, and that Y chromosome-specific repetitive sequences accumulated during its evolution. Together with the data from the genetic mapping of sex chromosome-linked genes, these results are very informative for understanding the structure of the sex chromosomes and the process of their evolution.

INTRODUCTION

Sex chromosomes evolved from a pair of autosomes, independently, in the various phyla of the plant kingdom at different times. After the appearance of the gene involved in heterogametic male (XY)-determination on the ancient Y chromosome, extensive recombination suppression between the ancient sex chromosomes evolved. Theoretically, the ancient Y chromosome then suffered a rapid accumulation of deleterious mutations and loss-of-function genes [1]. As a result, "gene deserts", or gene-poor heterochromatin regions became distributed over the Y chromosome. The gene-poor Y chromosomes are thought to be fated to become inert and therefore may disappear from the male genome at some point in the future.

Transposable elements (TEs) and non-coding repetitive sequences, has been widely noted as evolutionary forces that produce novel gene function and induce chromosome rearrangements and genome diversification [2]. Accumulation of these repetitive DNA sequences also contributes to the development of the gene deserts found in Y chromosomes. Therefore, it is possible that repetitive DNA accumulation starts at an early stage in the evolution of sex chromosomes [3]. Past cytogenetic analyses and recent genome projects have both revealed that, in animals, many Y chromosomes have more abundant heterochromatin derived from repetitive sequences than X chromosomes and autosomes. Accumulation of repetitive sequences induces abnormal recombination and chromosome breaks. Thus, repetitive DNA accumulation seems to be a plausible factor in the generation of the differences in morphology and size seen between X and Y chromosomes. For example, in both the fruit fly *Drosophila melanogaster* and in humans, the Y chromosome is drastically smaller than the X chromosome. The heterochromatic regions that are distributed over more than half of the human Y chromosome originated about 300 million years ago (mya) and the *D. melanogaster* Y chromosome, formed at least 60 mya, has become almost entirely heterochromatic. By contrast, the neo-Y chromosome of *D. miranda*, formed by a Y-autosome fusion only 1.2 mya, still harbors many functional genes. Even in this much younger Y chromosome, compared with the X chromosome, there is a more than 20-fold greater accumulation of repetitive sequences, mainly TEs [4]. These findings in animal species demonstrate that the accumulation of repetitive DNA is an important step in promoting the morphogenesis of sex chromosomes.

Repetitive DNA accumulation on Y chromosomes was believed to be a symptom of Y chromosome degeneration. The insertion of repetitive sequences into coding genes and regulatory regions induces alterations of gene function and results in gene loss. However, no correlation between TE insertion and gene dysfunction on the Y chromosome of *D. miranda* has been found [4]. Moreover, the rate of gene gain on the *Drosophila* Y chromosome is more than ten times the rate of gene loss, in contrast with the mammalian Y chromosome [5]. The contradiction between gene acquisition and the accumulation of highly repetitive sequences on the *Drosophila* Y chromosome, indicates that repetitive DNA accumulation is not always directly connected with Y chromosome degeneration.

To answer the question of how sex chromosomes are formed and evolve, we should survey more sex chromosomes in different taxa of plants. In contrast to animals, the majority of plant species do not have sex chromosomes, making dioecious species (with male and female functions on separate plants) a minority. However, plant sex chromosomes are found across the taxa from moss to flowering plants including familiar crop species like asparagus, hop, kiwifruit, papaya and spinach [6]. The ancient Y chromosome in the liverwort *Marchantia polymorpha* is small and largely heterochromatic [7], whereas in flowering plants most Y chromosomes are the largest chromosomes in the male genome and many plant sex chromosomes are morphologically indistinguishable from the autosomes [8, 9]. Why do the sex chromosomes in flowering plants seem to retain primitive characters similar to a pair of autosomes? One possible answer is that the plant sex chromosomes are evolutionarily very young. Thus, the study of sex chromosomes in flowering plants may allow us to catch a glimpse of the early steps in the genetic separation of male and female. It will, for example, give us the opportunity to study the problem of whether repetitive DNA accumulation occurs before or after gene degeneration on Y chromosomes [10].

Accumulation of repetitive sequences was generally found in the evolutionarily young plant Y chromosomes [11, 12]. The papaya Y chromosome is the youngest in plants, having diverged from the X chromosome only 2–3 mya [13]. Even in such young Y chromosomes, the accumulation of repetitive sequences and heterochromatinization can be detected. The white campion *Silene latifolia* is a flowering plant in which the sex chromosomes were first discovered in 1923; the Y and X chromosomes are the largest and second-largest chromosomes, respectively, of the male *S. latifolia*. The Y chromosome evolved in the *Silene* genus 10–20 mya through chromosomal inversion [14]. There is a large accumulation of many kinds of repetitive sequences on this young and large Y chromosome (See below).

The studies of the young Y chromosomes of *S. latifolia* and *D. miranda* strongly suggest that the accumulation of repetitive sequences is a crucial and common event in the early stages of the formation of both plant and animal sex chromosomes, although the level of accumulation and types of repetitive sequences vary. There is, however, another possible plant-specific function of repetitive sequences during sex chromosome evolution because, unlike in animals, plant genomes have increased through repetition by whole genome duplication with the amplification of repetitive sequences. Such repetitive DNA accumulation could contribute to increasing the size of plant Y chromosomes that maintain the largest amount of DNA in the male genome. Detailed investigation of the *S. latifolia* sex chromosome promises new insights into fundamental issues of the birth and evolution of sex chromosomes. In this chapter, we review the recent analyses by cytogenetic methods of repetitive sequences on the *S. latifolia* sex chromosomes.

FISH ANALYSIS OF REPETITIVE SEQUENCES ON THE SEX CHROMOSOMES OF *S. LATIFOLIA*

Repetitive Sequences Used for Fluorescence In Situ Hybridization (FISH) Analysis

Fluorescence *in situ* hybridization (FISH) has facilitated the investigation of sex chromosome structure and provided clues about sex-chromosome evolution [11]. In most FISH analyses, repetitive sequences have been used as probes, because their high copy numbers in plant genomes makes their FISH signals easy to detect. Repetitive sequences are classified mainly into dispersed DNAs and tandemly repeated DNAs [for reviews see 15–17]. Tandemly repeated DNAs are condensed in the centromeric or subtelomeric regions of the genome. In *S. latifolia*, the X43.1 family [11, 18–22], 15Ssp [23], and TR1 [24] were identified as subtelomeric tandem repeats, while STAR-C was characterized as a centromeric tandem repeat [24]. Dispersed DNAs include DNA transposon, retroelement and its remnant and each of them exclusively increases its copy number in plant genomes by individual replication strategies. Although dispersed DNAs are generally located in all regions of the genome, some dispersed DNAs are localized in particular chromosomal regions. In *S. latifolia*, the *Ty1-copia*-like retrotransposon (LAMR1) and *Ty3-gypsy*-like retrotransposon (LAGYP1) accumulate in all but the subtelomeric regions [25]. By contrast, another *gypsy*-like retrotransposon, *Retand*, is located mostly in the subtelomeric regions [26]. The proportion of these repetitive elements in the genome was systematically analyzed and it was found that the most abundant tandem repeat and dispersed repeat were the X43.1 family and the gypsy-like retrotransposons, respectively [24]. Distribution of these repetitive sequences on the sex chromosomes was well described previously [11, 12]. In addition to probes of repetitive sequences, genic probes [27–30] and BAC probes [31, 32] have also been used in FISH analyses. Multi-colored FISH using these probes allows FISH karyotyping to be conducted [31].

FISH probes can also be isolated by cytogenetic approaches regardless of the kind of repetitive sequences. Two main approaches, flow sorting and chromosome microdissection, have been developed [for a review, see 33]. Although the flow sorting approach would be effective in most species, it is not suitable for separation of the Y chromosome in *S. latifolia* because the Y chromosome is almost the same size as that of contiguous two autosomes and the Y chromosome fraction is contaminated with the autosomal clumps [34]. On the other hand, the chromosomal microdissection approach can visually separate a chromosome or a subchromosomal region regardless of chromosomal size [35]. The separated chromosome can be used for chromosome painting after amplification by the degenerate oligonucleotide-primed polymerase chain reaction (DOP-PCR). This strategy is commonly used in animal species [36], but is very difficult to apply directly in plants because of the wide spread of intermingled repetitive sequences [15, 19]. Hobza et al. [37] successfully performed sex chromosome painting in *S. latifolia* by modifying a FISH protocol using probes isolated by chromosomal microdissection (see below).

Rearrangement of Sex Chromosomes Suggested by Localization of Repetitive Sequences

FISH analysis using the X43.1 tandem repeat family clearly shows differences in the structure of the X and Y sex chromosomes. This repeat family has been used in many FISH analyses [11, 14, 18–22, 38, 39]. The FISH signals attributable to the X43.1 family were detected in the subtelomeric regions on both arms of the X chromosome, but only on the q arm of the Y chromosome where the pseudoautosomal region (PAR) is located. The regions that show the X43.1 signal correspond to the heterochromatic regions of the chromosomes as revealed by C-banding analysis [40]. These results suggest that the Y chromosome arm that does not show the X43.1 signal has different DNA components than the pseudo autosomal region (PAR).

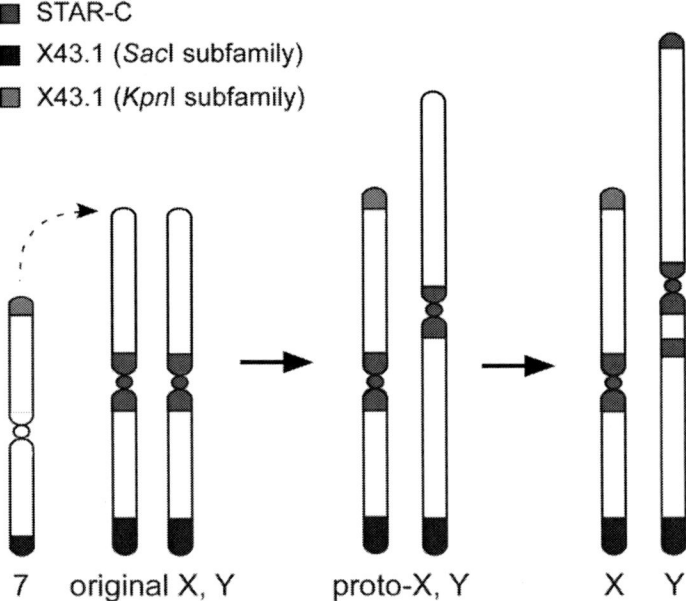

Figure. Schematic diagram of sex chromosome evolution in *S. latifolia* based on FISH data. The FISH signals of each probe are indicated by their respective colors. In the current model, the translocation of the *Kpn*I subfamily is thought to be involved in forming one end of the X chromosome q arm and the translocation of STAR-C may be involved in a large inversion in the Y chromosome. For the translocation of the *Kpn*I subfamily, an alternative scenario is possible; a translocation of the *Kpn*I subfamily from the original Y chromosome to chromosome 7. In this case, both original X and Y chromosomes should have the *Kpn*I subfamily on the opposite end from the *Sac*I subfamily localizing ends.

A detailed sequence analysis of the X43.1 repeat units provided knowledge of the sex chromosome structure and evolution. The X43.1 family was classified into four subfamilies comprising the *Sac*I subfamily (homologous to the originally isolated *Sac*I family [41]), *Kpn*I subfamily [21], #11F02 subfamily, and #16E07 subfamily. The mean inter-subfamily divergence between the *Sac*I and *Kpn*I subfamilies was 18.5%, and the mean intra-subfamily divergence was 9.1% for the *Sac*I subfamily and 10.1% for the *Kpn*I subfamily. The inter-subfamily divergences can be utilized for differential labeling of the chromosome ends. The bicolored FISH using the *Kpn*I and *Sac*I subfamilies as probes, showed specific signal intensities and locations at the chromosomal ends [22]. The signal color on the end of the Y chromosome q arm was similar to that on the X chromosome p arm, indicating that these chromosomal ends contain the same subfamily. These data are in good agreement with the model of the *S. latifolia* sex chromosome proposed by Lengerova et al. [28]. They suggested that the PAR on the X chromosome is located on the p arm. This is contrary to the classical model which locates the PAR on the q arm of the X chromosome. The signal color on the end of the q arm of the X chromosome was similar to that on the two ends of the autosomes. These ends possibly contain the *Kpn*I subfamily [21, 22]. BAC-FISH using a genic probe neighboring a cluster of the *Kpn*I subfamily showed signals on the end of the X chromosome q arm and on the end of chromosome 7. From these data, it has been postulated that the end of the X chromosome q arm may have recombined with the end of chromosome 7 (Figure) [42]. A gene translocation from an autosome to the X chromosome during sex chromosome evolution has also been reported for the X-linked gene, *Slcyt* [43]. Therefore, it is possible that such rearrangements may have occurred during the evolution of the sex chromosomes.

Indications of the presence of Y chromosome rearrangements have been found in studies of chromosomal interstitial telomeric repeats (ITRs). ITRs were identified in *Arabidopsis thaliana* and in *S. latifoia* [44, 45], and their adjacent sequences have been named ITR-adjacent sequences (IASs). None of the IASs were homologous to the isolated subtelomeric repetitive sequences of *S. latifolia*. One of the IASs, IAS-d, is distributed in the interstitial and proximal regions of the sex chromosomes, but not in the subtelomeric regions [45]. We postulate that IASs first accumulate in the subtelomeres of *S. latifolia* autosomes and then segmental regions of the subtelomere move into interstitial regions via chromosomal rearrangements. A similar result was found by FISH analysis using the STAR-Y probe [46]. STAR-Y is a short version of the centromeric repetitive sequence STAR-C. The FISH signals were detected exclusively on two parts of the Y chromosome; the central part of the q arm and the subtelomeric region of the p arm. This signal pattern is most likely a consequence of inversions on the Y chromosome (Figure). Together with other analysis, Hobza et al. proposed that at least two large inversions have contributed to Y chromosome evolution. This model was later modified by taking into account data from the deletion mapping of Y chromosome linked genes [47, 48].

Accumulation of Sex Chromosome-Specific Repetitive Sequences During Sex Chromosome Evolution

The evolution of structurally distinct sex chromosomes from a pair of autosomes is thought to involve a stepwise restriction of recombination between proto-Y and proto-X chromosomes [1, 49]. In this process, the accumulation of repetitive sequences such as STAR-Y might induce the increase in size of the sex chromosome that is observed in almost all heteromorphic Y chromosomes in the angiosperms. The accumulation of chromosome-specific repeat sequences was first clearly visualized by Hobza et al. [37]. They conducted modified FAST-FISH, in which a probe from microdissected chromosomes was hybridized using a short hybridization time and a low concentration of probes. The modified FAST-FISH successfully produced chromosome-specific signals. The microdissected X and Y probes showed signals along the entire X and Y chromosome, respectively, while signals on the other chromosomes were of lower intensity [37, 39]. Sex chromosome-specific distributions of some DNAs have been reported by several researchers. Kejnovsky et al. reported that the X and Y chromosomes show different signal patterns and that some of the interstitial signals are derived from chloroplast DNA [50]. Bergero et al. showed that one of the transposable elements, the miniature inverted repeat transposable element (MITE), had a high insertion rate on the Y chromosome [51]. These results imply that chromosome-specific invasion by these DNAs could drive the divergence of the sex chromosome sequences.

Another interesting recent finding is that the transposable element *SlOgre1* has accumulated all over the *S. latifolia* genome, but is absent in the Y chromosome [52]. The wide distribution of this transposable element in the genome was also found in three other dioecious *Silene* species from section Elisanthe, *S. dioica*, *S. diclinis*, and *S. marizii*. However, this element does not accumulate in the nondioecious species, *S. vulgaris*. These species are very closely related and are likely to have separated within the last 1–2 million years. It is probable that the spread of the *SlOgre1* element occurred before the sex chromosomes developed. Together with other analysis, Filatov et al. have suggested that the *SlOgre1* element might have spread after initiating the cessation of recombination between the X and Y chromosomes. Why the *SlOgre1* element is absent in the Y chromosome remains unknown. A possible explanation is that this element was activated only in female plants. However, the *SlOgre1* element is no longer transcribed and so no evidence has been obtained.

The *SlOgre1* element can be used for a "negative paint" of the Y chromosome. Using the negative paint, Howell et al. found that *S. diclinis* has neo-sex chromosomes composed of Y1, Y2, X, and neo-X [30]. Neo-sex chromosomes may be produced by reciprocal translocations between the original Y chromosome and an autosome. These four neo-sex chromosomes form a chain quadrivalent in male meiosis. This finding provides additional evidence for the dynamic rearrangement occurring in the sex chromosomes as described this section.

CONCLUSION

FISH analyses using repetitive sequences have contributed to our understanding of the rearrangement of sex chromosomes and the accumulation of the specific DNAs of sex chromosomes during their evolution. A detailed schematic map of the sex chromosomes including all FISH data was provided by Kejnovsky et al., [12]. Many repetitive sequences are distributed on the X and Y chromosomes and some of these accumulate predominantly on the Y chromosome. The accumulation of repetitive sequences can be explained by reduced selective pressure against insertion which contributes to the expansion of the sex chromosomes. However, the sequence and timing of these events is not well understood. To elucidate sex chromosome evolution more precisely, more detailed analyses of repetitive sequences are needed. For example, sequence divergence analysis and FISH analysis as described in Howell et al. [30] could be carried out in related dioecious and nondioecious species. Kejnovsky et al. (2010) performed 454 sequencing experiments to characterize repetitive sequences of *S. latifolia* and concluded that no other tandem repeats exist in the *S. latifolia* genome. [48]. Therefore, BAC-FISH analysis using non-repetitive regions of BAC clones as probes may be a powerful method for further studies.

Although mapping of the sex chromosome linked gene provided information for sex chromosome evolution, there are unexplained discrepancies between cytological data and genetic mapping data for the Y chromosome linked genes. Mapping of the sex chromosome linked genes indicated that more inversions than those suggested by the FISH data may occur during the construction of the Y chromosome [48, 53–55]. A new model including both FISH data and gene mapping data has been well described in Kejnovsky et al. [48]. However, this model does not explain the STAR-C location on the Y-chromosome p arm. More FISH experiments using Y-chromosome linked BAC clones will help improve the model and contribute to the elucidation of sex-chromosome evolution.

ACKNOWLEDGMENTS

The work in this chapter was supported by a Grant-in-Aid for Scientific Research (no. 20780009 to YK) from the Ministry of Education, Culture, Sports, Science, and Technology of Japan.

REFERENCES

[1]   Charlesworth D, Charlesworth B, Marais G: Steps in the evolution of heteromorphic sex chromosmes. *Heredity*, 2005, 95:118-128.

[2]   Biemont C, Vieira C: Genetics: junk DNA as an evolutionary force. *Nature*, 2006, 443:521-524.

[3]   Matsunaga S: Junk DNA promotes sex chromosome evolution. *Heredity*, 2009, 102:525-526.

[4]   Bachtrog D, Hom E, Wong KM, Maside X, de Jong P: Genomic degradation of a young Y chromosome in *Drosophila miranda*. *Genome Biol*, 2008, 9:R30.

[5]   Koerich LB, Wang X, Clark AG, Carvalho AB: Low conservation of gene content in the *Drosophila* Y chromosome. *Nature*, 2008, 456:949-951.

[6]   Matsunaga S, Kawano S: Sex determination by sex chromosomes in dioecious plants. *Plant Biology*, 2001, 3:481-488.

[7]   Yamato KT, Ishizaki K, Fujisawa M, Okada S, Nakayama S, Fujishita M et al.: Gene organization of the liverwort Y chromosome reveals distinct sex

chromosome evolution in a haploid system. *Proc Natl Acad Sci U S A*, 2007, 104:6472-6477.

[8]   Matsunaga S: Sex chromosome-linked genes in plants. *Genes Genet Syst*, 2006, 81: 219-226.

[9]   Jamilena M, Mariotti B, Manzano S: Plant sex chromosomes: molecular structure and function. *Cytogenet Genome Res*, 2008, 120:255-264.

[10]   Marais GA, Nicolas M, Bergero R, Chambrier P, Kejnovsky E, Moneger F *et al.*: Evidence for degeneration of the Y chromosome in the dioecious plant *Silene latifolia*. *Curr Biol*, 2008, 18:545-549.

[11]   Kazama Y, Matsunaga S: The use of repetitive DNA in cytogenetic studies of plant sex chromosomes. *Cytogenet Genome Res*, 2008, 120:247-254.

[12]   Kejnovsky E, Hobza R, Cermak T, Kubat Z, Vyskot B: The role of repetitive DNA in structure and evolution of sex chromosomes in plants. *Heredity*, 2009, 102:533-541.

[13]   Liu Z, Moore PH, Ma H, Ackerman CM, Ragiba M, Yu Q, Pearl HM, Kim MS, Charlton JW, Stiles JI, Zee FT, Paterson AH, Ming R: A primitive Y chromosome in papaya marks incipient sex chromosome evolution. *Nature*, 2004, 427:348-352.

[14]   Armstrong SJ, Filatov DA: A cytogenetic view of sex chromosome evolution in plants. *Cytogenet Genome Res*, 2008, 120:241-246.

[15]   Schmidt T, Heslop-Harrison JS: Genomes, genes and junk: the large-scale organization of plant chromosomes. *Trends Plant Sci*, 1998, 3:195-199.

[16]   Kumar A, Bennetzen JL: Plant retrotransposons. *Annu Rev Genet*, 1999. 33:479-532.

[17]   Heslop-Harrison JS: Comparative genome organization in plants: from sequence and markers to chromatin and chromosomes. *Plant Cell*, 2000, 12:617-635.

[18]   Buzek J, Koutnikova H, Houben A, Riha K, Janousek B, Siroky J, Grant S, Vyskot B: Isolation and characterization of X chromosome-derived DNA sequences from a dioecious plant *Melandrium album*. *Chromosome Res*, 1997, 5:57-65.

[19]   Matsunaga S, Kawano S, Michimoto T, Higashiyama T, Nakao S, Sakai A, Kuroiwa T: Semi-automatic laser beam microdissection of the Y chromosome and analysis of Y chromosome DNA in a dioecious plant, *Silene latifolia*. *Plant Cell Physiol*, 1999, 40:60-68.

[20]   Sykorova E, Fajkus J, Ito M, Fukui K: Transition between two forms of heterochromatin at plant subtelomeres. *Chromosome Res*, 2001, 9:309-323.

[21]   Kazama Y, Sugiyama R, Matsunaga S, Shibata F, Uchida W, Hizume M, Kawano S: Organization of the *KpnI* family of chromosomal distal-end satellite DNAs in *Silene latifolia*. *J Plant Res*, 2003, 116:317-326.

[22]   Kazama Y, Sugiyama R, Suto Y, Uchida W, Kawano S: The clustering of four subfamilies of satellite DNA at individual chromosome ends in *Silene latifolia*. *Genome*, 2006, 49:520-530.

[23]   Sykorova E, Cartagena J, Horakova M, Fukui K, Fajkus J. Characterization of telomere-subtelomere junctions in *Silene latifolia*. *Mol Gen Genomics*, 2003, 269:13-20.

[24]   Cermak T, Kubat Z, Hobza R, Koblizkova A, Widmer A, Macas J, Vyskot B, Kejnovsky E: Survey of repetitive sequences in *Silene latifolia* with respect to their distribution on sex chromosomes. *Chromosome Res*, 2008, 16:961-976.

[25]   Matsunaga S, Yagisawa F, Yamamoto M, Uchida W, Nakao S, Kawano S: LTR retrotranspsosns in the dioecious plant *Silene latifolia*. *Genome*, 2002, 45:745-751.

[26]   Kejnovsky E, Kubat Z, Macas J, Hobza R, Mracek J, Vyskot B: Retand: a novel family of gypsy-like retrotransposons harboring an amplified tandem repeat. *Mol Genet Genomics*, 2006, 276:254-263.

[27]   Moore RC, Kozyreva O, Lebel-Hardenack S, Siroky J, Hobza R, Vyskot B, Grant SR: Genetic and functional analysis of *DD44*, a sex-linked gene from the dioecious plant *Silene latifolia*, provides clues to early events in sex chromosome evolution. *Genetics*, 2003, 163:321-334.

[28]   Lengerova M, Moore RC, Grant SR, Vyskot B: The sex chromosomes of *Silene latifolia* revisited and revised. *Genetics*, 2003, 165:935-938.

[29]   Fujita N, Ishii K, Kawano S. An STS Marker, Y202, Located on the *Silene latifolia* Y Chromosome between the Chromosomal Distal-end Satellite DNA and *SlY1*. *Cytologia*, 2008, 73: 159-165.

[30]   Howell EC, Armstrong SJ, Filatov DA. Evolution of neo-sex chromosomes in *Silene diclinis*. *Genetics*, 2009, 182:1109-1115.

[31]   Lengerova M, Kejnovsky E, Hobza R, Macas J, Grant SR, Vyskot B: Multicolor FISH mapping of the dioecious model plant, *Silene latifolia*. *Theor Appl Genet*, 2004, 108:1193-1199.

[32]   Ishii K, Sugiyama R, Onuki M, Kazama Y, Matsunaga S, Kawano S: The Y chromosome-specific STS marker MS2 and its peripheral regions on the

Y chromosome of the dioecious plant *Silene latifolia*. *Genome*, 2008, 51:251-260.

[33]   Gill BS, Friebe B: Plant cytogenetics at the dawn of the 21st century. *Curr Opin Plant Biol*, 1998, 1:109-115.

[34]   Kejnovsky E, Vrana J, Matsunaga S, Soucek P, Siroky J, Dolezel J, and Vyskot B: Localization of male-specifically expressed *MROS* genes of *Silene latifolia* by PCR on flow-sorted sex chromosomes and autosomes. *Genetics*, 2001, 158:1269-1277.

[35]   Meltzer PS, Guan XY, Burgess A, Trent JM: Rapid generation of region specific probes by chromosome microdissection and their application. *Nat Genet*, 1992, 1:24-28.

[36]   Ried T, Schrock E, Ning Y, Wienberg J: Chromosome painting: a useful art. *Hum Mol Genet*, 1998, 7:1619-1626.

[37]   Hobza R, Lengerova M, Cernohorska H, Rubes J, Vyskot B: FAST-FISH with laser beam microdissected DOP-PCR probe distinguishes the sex chromosome of *Silene latifolia*. *Chromosome Res*, 2004, 12:245-250.

[38]   Hobza R, Lengerova M, Svoboda J, Kubekova H, Kejnovsky E, Vyskot B: An accumulation of tandem DNA repeats on the Y chromosome in *Silene latifolia* during early stages of sex chromosome evolution. *Chromosoma*, 2006, 115:376-382.

[39]   Hobza R, Vyskot B: Laser microdissection-based analysis of plant sex chromosomes. *Methods Cell Biol*, 2007, 82:433-453.

[40]   Grabowska-Joachimiak A, Joachimiak A: C-banded karyotypes of two *Silene* species with heteromorphic sex chromosomes. *Genome*, 2002, 45:243-252.

[41]   Garrido-Romos MA, de la Herran R, Ruiz Rejon M, Ruiz Rejon C: A subtelomeric satellite DNA family isolated from the genome of the dioecious plant *Silene latifolia*. *Genome*, 1999, 42:442-446.

[42]   Ishii K, Amanai Y, Kazama Y, Ikeda M, Kamada H, Kawano S. Analysis of BAC clones containing homologous sequences on the end of the Xq arm and on chromosome 7 in the dioecious plant *Silene latifolia*. *Genome*, 2010, 53:311-320.

[43]   Kaiser VB, Bergero R, Charlesworth D: *Slcyt*, a newly identified sex-linked gene, has recently moved onto the X chromosome in *Silene latifolia* (Caryophyllaceae). *Mol Biol Evol*, 2009, 26:2343-2351.

[44]   Uchida W, Matsunaga S, Sugiyama R, Kawano S: Interstitial telomere-like repeats in the *Arabidopsis thaliana* genome. *Genes Genet Syst*, 2002, 77:63-67.

[45]   Uchida W, Matsunaga S, Sugiyama R, Shibata F, Kazama Y, Miyazawa Y, Hizume M, Kawano S: Distribution of interstitial telomere-like repeats and their adjacent sequences in a dioecious plant, *Silene latifolia*. *Chromosoma*, 2002, 111:313-320.

[46]   Hobza R, Kejnovsky E, Vyskot B, Widmer A. The role of chromosomal rearrangements in the evolution of *Silene latifolia* sex chromosomes. *Mol Genet Genomics*, 2007, 278: 633-638.

[47]   Bergero R, Charlesworth D, Filatov DA, Moore RC: Defining regions and rearrangements of the *Silene latifolia* Y chromosome. *Genetics*, 2008, 178:2045-2053.

[48]   Kejnovsky E, Vyskot B: *Silene latifolia*: The classical model to study heteromorphic sex chromosomes. *Cytogenet Genome Res*, 2010, 129: 250-262.

[49]   Charlesworth B: The evolution of chromosomal sex determination and dosage compensation. *Curr Biol*, 1996, 6:149-162.

[50]   Kejnovsky E, Kubat Z, Hobza R, Lengerova M, Sato S, Tabata S, Fukui K, Matsunaga S, Vyskot B: Accumulation of chloroplast DNA sequences on the Y chromosome of *Silene latifolia*. *Genetica*, 2006, 128:167-175.

[51]   Bergero R, Forrest A, Charlesworth D: Active miniature transposons from a plant genome and its nonrecombining Y chromosome. *Genetics*, 2008, 178:1085-1092.

[52]   Filatov DA, Howell EC, Groutides C, Armstrong SJ: Recent spread of a retrotransposon in the *Silene latifolia* genome, apart from the Y chromosome. *Genetics*, 2009, 181:811-817.

[53]   Nicolas M, Marais G, Hykelova V, Janousek B, Laporte V, Vyskot B, Mouchiroud D, Negrutiu I, Charlesworth D, Moneger F: A Gradual process of recombination restriction in the evolutionary history of the sex chromosomes in dioecious plants. *PLoS Biol*, 2005, 3:e4.

[54]   Zluvova J, Janousek B, Negrutiu I, Vyskot B: Comparison of the X and Y chromosome organization in *Silene latifolia*. *Genetics*, 2005, 170:1431-1434.

[55]   Zluvova J, Georgiev S, Janousek B, Charlesworth D, Vyskot B, Negrutiu I: Early events in the evolution of the *Silene latifolia* Y chromosome: male specialization and recombination arrest. *Genetics*, 2007, 177:375-386.

In: New Insights on Plant Sex Chromosomes   ISBN: 978-1-61470-236-8
Editor: Rafael Navajas-Pérez   ©2012 Nova Science Publishers, Inc.

Chapter III

How Much Do We Know about Evolution of Sex Chromosomes in Plants?

Dmitry A. Filatov

Department of Plant Sciences, University of Oxford, Oxford, UK

According to the currently accepted model, sex chromosomes evolve from a pair of autosomes via a series of step-wise expansions of the non-recombining region on the Y chromosome (NRY), creating 'evolutionary strata', followed by degeneration of genes trapped in the NRY. This model is based on the analysis of ancient (over 200 million year [MY] old) sex chromosomes of mammals and birds, as well as recent translocations of autosomes to sex chromosomes (neo-sex chromosomes) in *Drosophila*. Much less is known about plant sex chromosomes, with the bulk of data coming from only two species, *Silene latifolia* and *Carica papaya*, and even for these best studied plants, less than a dozen sex-linked genes have been identified and analysed. Here I critically discuss the main conclusions from the data available. In particular, I stress that the evidence for 'evolutionary strata' and for genetic degeneration of Y-linked genes in plants is relatively weak. Thus, it is premature to extend the 'animal model' of sex chromosome evolution to plants, as the evolutionary trajectory of plant sex chromosomes may differ from that in animals.

Sex chromosomes are peculiar structures present in most animal and some plant genomes [1]. Although they evolved many times independently, their properties are quite similar in different groups of organisms. The chromosome restricted to one sex (Y in male- or W in female-heterogametic species) lacks recombination and is genetically degenerate, while its counterpart (X or Z, respectively) is present in both sexes, actively recombines in the homogametic sex and is relatively gene-rich. Independent evolution of similar structures in different organisms provides a wealth of material for evolutionary genetic studies seeking to understand the origin and evolutionary forces involved in sex chromosome evolution.

SEX CHROMOSOME EVOLUTION IN ANIMALS AND PLANTS

Significant progress in understanding sex chromosome evolution has been achieved with the availability of genome sequences of multiple animal species [2-5]. Despite the ancient origin of sex chromosomes in mammals, parts of the X- and Y-chromosomes are still homologous to each other, suggesting that they evolved from a pair of autosomes. Comparisons of sequences of sex chromosomes of different mammalian species revealed a dynamic picture of multiple translocations of autosomal material to sex chromosomes as well as a series of expansions of the non-recombining region (NRY) [6, 7]. Human sex chromosomes contain recognisable traces of at least four expansions of the NRY, termed 'evolutionary strata' [8]. The most recent 'stratum' that was added to the primate NRY ca. 20 million years ago. It is still a part of a pseudo-autosomal region in cows [9] and is gene-rich. All but four of the genes in this region have been lost from the human Y-chromosome. Such gene loss appears to be a general property of the Y-chromosomes and there are theoretical reasons to expect genetic degeneration of non-recombining regions [10].

Complete linkage of mutations in non-recombining regions renders genes evolutionary non-independent and selection at one locus can affect evolutionary fate of the other locus [11]. This causes 'evolutionary interference' of mutations that reduce efficacy of natural selection, leading to higher chance of fixing detrimental mutations (e.g. gene loss) and lower chance for advantageous mutations to spread [12]. For example, elimination of strongly deleterious mutations from the population reduces the pool of chromosomes contributing to the next generation, which inflates stochastic fixation of weakly deleterious mutations [13]. This process is referred to as background selection. On the other hand, a strong advantageous mutation rising in frequency due to positive selection might drag along linked weakly deleterious mutations [14]. This process is often referred to as hitchhiking or selective sweeps. For example, fixation of a truncated copy of *USP9Y* gene and shortening of *TBL1Y* on the chimpanzee Y-chromosome is thought to be due to strong sperm competition in this promiscuous species [15]. Several genes on the Y-chromosome are involved in spermatogenesis, and mutations creating advantage in sperm competition should be strongly advantageous in promiscuous species. In contrast to chimpanzees, somewhat less promiscuous *Homo sapiens* lost no genes from the Y chromosome since divergence from chimpanzees [15]. Although the loss of genes from the chimp Y-chromosome and no loss in humans may have occurred by chance, this example illustrates the principle of Y-degeneration due to hitchhiking.

Although there is little doubt that evolutionary interference and gene loss do occur in non-recombining regions, the particulars of the population genetic processes involved are still actively debated [12]. Plant sex chromosomes should be especially useful to study population genetic processes driving Y-chromosome degeneration. They are often of recent origin (see below) and probably contain many genes; thus, more evolutionary interference between the genes is expected to occur on plant Y chromosomes compared to their older, more degenerate, animal counterparts. Furthermore, the processes leading to genetic degeneration might differ between animals and plants and by comparing them we may better understand the causes of genetic degeneration of Y-linked genes. Loss of Y-linked genes in animals might be caused by lack of recombination on the Y, leading to 'evolutionary interference' between mutations; however it may also occur due to 'sheltering' of Y-linked genes by X-linked homologs resulting in relaxed purifying selection [16]. This is not the case in plants, as a significant proportion of a plant genome is expressed in the haploid state in pollen [17]. For example, in *Arabidopsis thaliana* 13996 out of 22592 genes were demonstrated to be actively expressed in pollen [18]. Such widespread haploid expression should result in exposure of recessive mutations to selection, which might prevent or slow down accumulation of deleterious mutations. If plant Y-chromosomes undergo degeneration, it provides a direct test for the evolutionary consequences of suppressed recombination, as the 'sheltering' effect is not present or less pronounced in plants.

Loss of genes from the Y-chromosome results in imbalance in expression of sex-linked genes between males and females: females have two functional X-linked copies, while males have only one copy. To correct for this imbalance different organisms have evolved different dosage compensation mechanisms –

mammals shut down expression of one of the X-chromosomes in females, while *Drosophila* increases expression from a single X-chromosome in males. It remains unclear how such compensating mechanisms evolve and expression analysis of X-linked genes in plants may help to resolve this issue. A particularly important aspect of plant sex chromosomes is their recent evolution *de novo*, in contrast to actively studied neo-sex chromosomes in *Drosophila miranda* [19]. Neo-sex chromosomes evolve due to a translocation of an autosome to old sex-chromosomes and evolution of neo-sex chromosomes may not follow exactly the same path as the sex chromosomes evolving *de novo*. For example, neo-sex chromosomes may 'inherit' dosage compensation system from the older pair of sex chromosomes, creating an impression of rapid evolution of dosage compensation, while sex chromosomes evolving *de novo* may take a long time to acquire dosage compensation.

SEX CHROMOSOMES IN PLANTS

Only about 6% of plant species are dioecious (i.e. have separate male and female individuals) [20] and only a few of these dioecious plants are known to have sex chromosomes [21, 22]. Plant sex chromosomes that are clearly distinguishable under a microscope have been reported only for *Silene latifolia* [22], *Rumex* [23], *Cannabis* [24], *Humulus* [25], and *Marchantia polymorpha* [26]. Among these plants *S. latifolia* represents the best studied example of heteromorphic sex chromosomes [27]. *Rumex* may be equally informative and interesting genus to study sex chromosomes [28], however there is very little molecular genetic data available at the moment. Sex chromosomes in most other studied dioecious plants are either absent, or present in a cryptic form, when X and Y look identical to each other, which is best studied in *Carica papaya* [29, 30].

Similar to animals, male heterogamety is more common among the plants with sex chromosomes (e.g. see table 1 in [21]). Both *S. latifolia* and *C. papaya* have male heterogamety, with XX-females and XY-males. Also both species evolved sex chromosomes from a pair of autosomes, as evidenced from the observation that X- and Y-chromosomes share homologous genes [31, 32]. For *Silene* it was demonstrated that homologs of the sex-linked genes map to the same chromosome in a non-dioecious relative, providing evidence for the origin of sex chromosomes from a single pair of autosomes [32].

Sex chromosomes in *S. latifolia* are the largest in the genome, with the Y about 1.4 times larger than the X [27]. X- and Y- chromosomes in *S. latifolia* pair in a single small pseudoautosomal region (PAR) at one end [33]. The presence of a second PAR in *Silene*, suggested by Scotti and Delph [34], has not been confirmed by a recent detailed cytogenetic analysis [35]. In contrast to *S. latifolia*, papaya sex-determining chromosomes are mainly composed of two large PARs with a small 'sex chromosomal' region about seven megabases long, where recombination is suppressed in males [29, 36]. Such a difference between the sex chromosomes of the two species probably reflects different stages of sex chromosome evolution, with papaya (and probably many other dioecious plants without heteromorphic sex chromosomes) at a relatively early stage of sex chromosome evolution and *S. latifolia* at a more advanced stage. Sex chromosomes in *Rumex* may be at an even more advanced stage, as the Y in *Rumex* is mainly heterochromatic [37], not dissimilar to old sex chromosomes in animals, while the *S. latifolia* Y-chromosome is mainly euchromatic [38].

HOW OLD ARE PLANT SEX CHROMOSOMES?

The absolute age of the sex chromosomes is difficult to assess precisely. In the absence of dated fossil evidence one has to rely on molecular clock to estimate the age. Generally, silent nucleotide divergence between the homologous X- and Y-linked genes is expected to be proportional to the time of divergence between the X- and Y-chromosomes. However, X/Y divergence varies widely between the genes. For example in *S. latifolia*, it ranges from less than 2% to over 30%. The highest X/Y divergence in *S. latifolia*, 35%, was reported between *MROS3X* gene and its degenerate homologue on the Y-chromosome [39]. However, this figure grossly exceeds the divergence between *S. latifolia* and its non-dioecious relatives (e.g, *S. vulgaris*), while all available evidence suggests that sex chromosomes evolved in *S. latifolia* ancestor after the split from other non-dioecious lineages. Thus, *MROS3X/Y* divergence most likely leads to an overestimation of maximal age of Silene sex chromosomes, possibly due to paralogy of the *MROS3X* and *MROS3Y* genes analysed by [39].

A more realistic estimate for the age of sex chromosome evolution in *S. latifolia* is provided by several genes with silent X/Y divergence of ~20% [40]. Assuming a substitution rate of $\sim 10^{-8}$ mutations per nucleotide per generation one can estimate the age of sex chromosomes in *S. latifolia* as ~10 million years

(MY), which is a widely quoted number for this species [41]. However, there are many estimates of mutation rate that can differ by an order of magnitude [42, 43]. The rate of the molecular clock is affected by many factors [44]. This issue must be particularly complex in plants, where there is no germline separated from somatic tissues. In plants germinal tissue is formed from somatic cells, thus the number of cell divisions between the seed and the flower depends on the size of a plant. With more cell divisions, higher per-generation mutation rate is expected in larger plants. In fact the situation is even more complex; the rate of molecular clock in plants is strongly dependent on life history and there is a great variation in substitution rates across different plant groups [45].

With no estimates of molecular clock rate in plant species with sex chromosomes, such as *S.latifolia* or *C.papaya*, one has to rely of the rates estimated in other organisms, which is problematic as molecular clock rate varies among the species. Thus, the widely quoted ~10 MY [41] or ~20MY [46] are no more than convenient round numbers used as 'estimates' for the age of *S. latifolia* sex chromosomes. Similar arguments with regard to dating of the origin of sex chromosomes apply to other plants. In particular, the estimate of 2-3 MY for papaya sex chromosomes [31, 47] has to be taken with caution, as in fact it may be several times higher or lower. Nevertheless, it is possible to compare the ages of sex chromosomes in terms of nucleotide divergence rather than years or generations. Such a comparison between *S. latifolia* and papaya indicates a more recent origin for papaya sex chromosomes [31, 40, 48]. Using the same approach for comparisons between plants and animals (e.g. mammals, birds) demonstrates much more recent origin for the plant sex chromosomes.

Why are plant sex chromosomes younger than in animals? There are many possible reasons for this apparent disparity. Firstly, this may partly be the matter of perception rather than the real difference. The majority of data about animal sex chromosomes comes from mammals and *Drosophila*, which do have ancient sex chromosomes. Much less is known about sex chromosomes of other animal groups. Amphibians and reptilians do have a wide diversity of sex-determining mechanisms, including temperature-dependent sex determination. Sex chromosomes, if present in species of these groups, are often very young [49, 50]. Sex determination in insects also varies between the groups: although most orders seem to have male heterogamety; Lepidoptera are female-heterogametic and Hymenoptera are haplo/diploid. On the other hand, when one discusses plant sex chromosomes, it is usually flowering plants that come to mind and other groups are rarely mentioned. This may significantly bias the picture, as angiosperms are of relatively recent origin and if the common ancestor did not have sex chromosomes, then any sex chromosomes that evolved within the group would be relatively young.

Another possible reason for younger sex chromosomes in plants is that dioecy is a mechanism for avoiding self-fertilisation and inbreeding, while angiosperms have alternative solutions to this problem. Self-incompatibility is an effective way to avoid selfing and it is fairly widespread in angiosperms, which makes dioecy and sex chromosomes redundant. In plants beyond angiosperms dioecy is fairly common. For example, many gymnosperm species are dioecious, such as Ginkgo and Cycads, though none of the gymnosperms have been confirmed to have sex chromosomes so far. *Ginkgo* was suggested to have sex chromosomes based on cytogenetic observations [51-53], however, the results of these studies are contradictory and the question requires further investigation. The best studied case of sex chromosomes beyond flowering plants is from investigations of the liverwort *Marchantia polymorpha*. The Y chromosome of this species has been sequenced, but unfortunately, no sequence of the X-chromosome is available, so it is difficult to assess how old sex chromosomes are in this species.

DO PLANTS FOLLOW THE ADDITION/ATTRITION MODEL OF SEX CHROMOSOME EVOLUTION?

There is now good evidence for multiple consecutive translocations of autosomal material to mammalian sex chromosomes (reviewed in [7]). Once the genetic material ends up in the non-recombining region, it starts to accumulate deleterious mutations and lose genes. This process of translocations and genetic degeneration is often referred to as 'addition/attrition' [54]. Does it also occur in plants?

If a translocation adds material to the non-PAR end of the X or the Y, then this results in formation of so-called neo-sex chromosomes, with more than one X- and/or Y-chromosome. For example, platypus and echidna have five X- and five Y-chromosomes that evolved via many consecutive translocations to the sex and neo-sex chromosomes [55, 56]. Similar translocation to sex chromosome has recently been reported in the close relative of *S. latifolia* – *Silene diclinis* [33]. This translocation resulted in the formation of neo-sex chromosomes that form a chain in male meiosis, similar to monotremes.

On the other hand, a translocation to the PAR expands this region on both the X- and Y-chromosomes because recombination in the PAR redistributes the translocation between the X- and Y-chromosomes. Such expanded PAR may then contract if a part of it is included into the non-recombining region. It is easy to imagine this process happening due to inversions on the Y chromosome [57]. An inversion including a part of the PAR and a part of the NRY will suppress recombination in the PAR region included in the inversion. This occurred several times in the history of human sex chromosomes, creating distinct regions of similar X/Y divergence called 'evolutionary strata' [8].

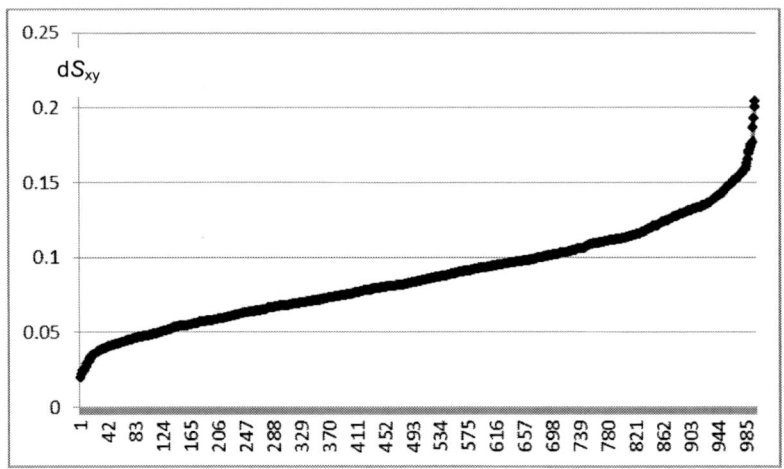

Figure 1. Simulated distribution of synonymous divergence between homologous X- and Y-linked genes (dS_{xy}, Y-axis) for sex chromosomes, where recombination has ceased at the same time across the entire length of the sex chromosomes (i.e. no 'evolutionary strata'). The genes (along the X-axis) are ranked according to their dS_{xy}. Coding sequences for X- and Y-linked copies of 1000 genes 200 codons long were simulated using evolver software from PAML package [59] with expected $dS_{xy} = 10\%$. Synonymous divergence was estimated using yn00 program [60] from the same package. All the variation between the genes in this simulation is purely due to stochastic accumulation of mutations and not due to differences in time since cessation of X/Y recombination. The range of variation in dS_{xy} is comparable to that observed on *S. latifolia* sex chromosomes. Thus, it may not be necessary to imply three separate cessations of recombination on *S. latifolia* sex chromosome to explain the variation in X/Y divergence.

The published sex-linked genes in *S. latifolia* were reported to fall into at least three groups, with X/Y silent divergence of 2-4%, 5-8% and >14%, resembling human 'evolutionary strata'. The difference in X/Y divergence is hypothesised to arise due to several steps of expansion of the non-recombining region [41], though no correspondence between inversions on the Y and the putative 'evolutionary strata' was found in *Silene* [58]. The evidence for evolutionary strata in *Silene* is based on only 10 genes and with more genes we might identify more groups of genes with similar X/Y divergence. Our unpublished data based on many more *S. latifolia* sex-linked genes isolated recently demonstrates that the distribution of X/Y divergence in this species is continuous (Filatov in prep). This might suggest a gradual, rather than stepwise, expansion of the non-recombining region. Alternatively, it is possible that all the variation in X/Y divergence among the genes is due to stochastic variation in the numbers of mutations accumulated in different genes and it is not necessary to postulate several NRY expansion steps. Indeed, a simple simulation shown in figure 1 demonstrates that a single event creating a non-recombining region along the entire chromosome can explain all the variation in X/Y divergence observed in *S. latifolia* and three separate expansions are not necessary to explain the observed picture. Sparse sampling of only a few genes from this distribution can create clusters of genes with similar divergence and lead to (possibly false) conclusion of 'evolutionary strata' present on the sex chromosomes of this species.

To resolve the question with 'evolutionary strata' in *S. latifolia* (and other species) many more than 10 genes with homologous X- and Y-linked copies will have to be isolated and analysed. This will allow one to test 1) whether the range of X/Y divergence is incompatible with purely stochastic variation of this parameter among the genes, 2) whether there are discontinuities in the distribution of X/Y divergence, and 3) whether the order of genes on the X chromosome corresponds to that expected from divergence. Until more *S. latifolia* sex-linked genes are available, the question of the presence of 'evolutionary strata' in this species will remain unresolved.

Regardless of whether the expansion of the NRY in *S. latifolia* occurred in one or several steps, it is clear that it contains many functional genes. Hence the question: do they undergo degeneration? The evidence for genetic degeneration of plant sex chromosomes is suggestive, but not conclusive.

The sequenced *Marchantia* Y chromosome contained only 64 functional genes in ~10 Mb of DNA [26, 61]. Gene density of ~6 genes per megabase is not particularly low and is similar to that in autosomal regions of the human genome.

The sequence of the *Marchantia* X-chromosome is not yet available, but PCR-amplification of Y-linked genes from females (containing no Y) revealed X-linked homologs for most of the Y-linked genes [61]. Thus, most *Marchantia* sex-linked genes appear intact and there is little evidence for sex chromosome degeneration in this species.

The evidence for genetic degeneration of Y-linked genes in *S. latifolia* is also inconclusive. There are signs of genetic degeneration in DNA polymorphism [62-64] and divergence [65, 66], however all but one gene isolated from *S. latifolia* X-chromosome appear to have apparently functional Y-linked homologs. The only gene that was reported to have a dysfunctional copy on the Y [39] is a member of multigene family [67] and is most likely a paralog rather than orthologous degenerate copy.

The lack of clear-cut evidence of genetic degeneration of the *S. latifolia* Y chromosome is somewhat surprising: the rate of Y chromosome degeneration depends on the number of functional genes linked together; the more functional genes linked on a non-recombining Y chromosome, the faster degeneration process should proceed. Given the size of *S. latifolia* sex chromosomes and their recent origin, the gene loss should be fairly rapid. Yet, we see little evidence for gene loss in *Silene*.

If plant Y chromosomes do undergo degeneration, this process is probably slower than in animals. The age of the *S. latifolia* Y-chromosome is comparable to the youngest region on the human Y chromosome, stratum 4 – synonymous X/Y divergence in human stratum 4 ranges from 5 to 12% [8], and all but one *S. latifolia* genes published so far either fall in this range or have even higher X/Y divergence. Yet, the human stratum 4 is almost completely degenerate, while all known *S. latifolia* Y-linked genes are still functional. Furthermore, more than half of the genes on the *D. miranda* neo-Y chromosome accumulated stop codons and lost function in only ~1.5 million years [68]. Although *S. latifolia* Y-chromosome is older than the neo-Y in *D. miranda*, no premature stop codons have been found in *S. latifolia* genes. It seems likely that slower loss of genes from plant Y-chromosomes is due to widespread haploid expression of plant genes [18]. However, this hypothesis remains to be tested. If confirmed, it will suggest that Y chromosome degeneration in animals mainly occurs due to recessive deleterious mutations and 'sheltering' of Y-linked genes by functional X-linked homologs plays a significant role in Y-degeneration.

CONCLUDING REMARKS

Although sex chromosomes perform same sex-determining function and generally, their evolution is probably driven by similar processes, there appear to be significant differences in sex chromosome evolution in animals and plants. These differences provide different perspectives on sex chromosome evolution, which may help us better understand the general evolutionary processes involved. However the lack of genes isolated from plant sex chromosomes and paucity of genomic information in plants remain significant obstacles for the progress in this field.

ACKNOWLEDGMENTS

I'm grateful to Mark Chapman and Graham Muir for corrections and suggestions that helped to improve the manuscript.

REFERENCES

1.   Bull, J.J., *Evolution of sex determining mechanisms.* 1983: The Benjamin/Cummings Publishing.
2.   Ming, R. and P.H. Moore, *Genomics of sex chromosomes.* Curr Opin Plant Biol, 2007. **10**(2): p. 123-30.
3.   Bellott, D.W. and D.C. Page, *Reconstructing the evolution of vertebrate sex chromosomes.* Cold Spring Harb Symp Quant Biol, 2009. **74**: p. 345-53.
4.   Hughes, J.F., et al., *Chimpanzee and human Y chromosomes are remarkably divergent in structure and gene content.* Nature, 2010. **463**(7280): p. 536-9.
5.   Carvalho, A.B., L.B. Koerich, and A.G. Clark, *Origin and evolution of Y chromosomes: Drosophila tales.* Trends Genet, 2009. **25**(6): p. 270-7.
6.   Marshall Graves, J.A., *Weird animal genomes and the evolution of vertebrate sex and sex chromosomes.* Annu Rev Genet, 2008. **42**: p. 565-86.
7.   Wilson, M.A. and K.D. Makova, *Genomic analyses of sex chromosome evolution.* Annu Rev Genomics Hum Genet, 2009. **10**: p. 333-54.
8.   Lahn, B.T. and D.C. Page, *Four evolutionary strata on the human X chromosome.* Science, 1999. **286**(5441): p. 964-7.

9. Van Laere, A., W. Coppieters, and M. Georges, *Characterization of the bovine pseudoautosomal boundary: documenting the evolutionary history of mammalian sex chromosomes.* Genome Res., 2008. **18**: p. 1884-1895.

10. Charlesworth, B. and D. Charlesworth, *The degeneration of Y chromosomes.* Philos Trans R Soc Lond B Biol Sci, 2000. **355**(1403): p. 1563-72.

11. Hill, W.G. and A. Robertson, *The effect of linkage on limits to artificial selection.* Genet Res, 1966. **8**(3): p. 269-94.

12. Charlesworth, D., *Sex chromosome origins and evolution*, in *Evolutionary Genomics and Proteomics*, P.A. Pagel M, Editor. 2008, Sinauer Associates: Sunderland. p. 207-240.

13. Charlesworth, B., *The effect of background selection against deleterious mutations on weakly selected, linked variants.* Genet Res, 1994. **63**(3): p. 213-27.

14. Rice, W.R., *Genetic hitchhiking and the evolution of reduced genetic activity of the Y sex chromosome.* Genetics, 1987. **116**(1): p. 161-7.

15. Hughes, J.F., et al., *Conservation of Y-linked genes during human evolution revealed by comparative sequencing in chimpanzee.* Nature, 2005. **437**: p. 100.

16. Nei, M., *Accumulation of non-functional genes on sheltered chromosomes.* American Naturalist, 1970. **104**: p. 311-322.

17. Borg, M., L. Brownfield, and D. Twell, *Male gametophyte development: a molecular perspective.* J Exp Bot, 2009. **60**(5): p. 1465-78.

18. Honys, D. and D. Twell, *Transcriptome analysis of haploid male gametophyte development in Arabidopsis.* Genome Biol, 2004. **5**(11): p. R85.

19. Bachtrog, D., N.R. Toda, and S. Lockton, *Dosage compensation and demasculinization of X chromosomes in Drosophila.* Curr Biol, 2010. **20**(16): p. 1476-81.

20. Renner, S.S. and R.E. Rickelfs, *Dioecy and its correlates in the flowering plants.* Am J Bot, 1995. **82**: p. 596-606.

21. Jamilena, M., B. Mariotti, and S. Manzano, *Plant sex chromosomes: molecular structure and function.* Cytogenet Genome Res, 2008. **120**(3-4): p. 255-64.

22. Westergaard, M., *The mechanism of sex determination in dioecious flowering plants.* Adv Genet, 1958. **9**: p. 217-81.

23. Navajas-Perez, R., et al., *Molecular cytogenetic characterization of Rumex papillaris, a dioecious plant with an XX/XY(1)Y (2) sex chromosome system.* Genetica, 2009. **135**(1): p. 87-93.

24. Sakamoto, K., et al., *Characterization, genome sizes and morphology of sex chromosomes in hemp (Cannabis sativa L.).* Cytologia, 1998. **63**: p. 459-464.

25. Grabowska-Joachimiak, A., et al., *C-Banding/DAPI and in situ hybridization reflect karyotype structure and sex chromosome differentiation in Humulus japonicus Siebold & Zucc.* Cytogenet Genome Res, 2011. **132**(3): p. 203-11.

26. Okada, S., et al., *The Y chromosome in the liverwort Marchantia polymorpha has accumulated unique repeat sequences harboring a male-specific gene.* Proc Natl Acad Sci U S A, 2001. **98**(16): p. 9454-9.

27. Armstrong, S.J. and D.A. Filatov, *A cytogenetic view of sex chromosome evolution in plants.* Cytogenet Genome Res, 2008. **120**(3-4): p. 241-6.

28. Cunado, N., et al., *The evolution of sex chromosomes in the genus Rumex (Polygonaceae): Identification of a new species with heteromorphic sex chromosomes.* Chromosome Res, 2007. **15**(7): p. 825-33.

29. Liu, Z., et al., *A primitive Y chromosome in papaya marks incipient sex chromosome evolution.* Nature, 2004. **427**(6972): p. 348-52.

30. Ming, R., Q. Yu, and P.H. Moore, *Sex determination in papaya.* Semin Cell Dev Biol, 2007. **18**(3): p. 401-8.

31. Yu, Q., et al., *Low X/Y divergence in four pairs of papaya sex-linked genes.* Plant J, 2008. **53**(1): p. 124-32.

32. Filatov, D.A., *Evolutionary history of Silene latifolia sex chromosomes revealed by genetic mapping of four genes.* Genetics, 2005. **170**(2): p. 975-9.

33. Howell, E.C., S.J. Armstrong, and D.A. Filatov, *Evolution of Neo-sex Chromosomes in Silene diclinis.* Genetics, 2009. **182**: p. 1109-1115.

34. Scotti, I. and L.F. Delph, *Selective trade-offs and sex-chromosome evolution in Silene latifolia.* Evolution, 2006. **60**(9): p. 1793-800.

35. Howell, E.C., S.J. Armstrong, and D.A. Filatov, *Dynamic gene order on the Silene latifolia Y chromosome.* Chromosoma 2011. **120**(3): p. 287-296.

36. Yu, Q., et al., *A physical map of the papaya genome with integrated genetic map and genome sequence.* BMC Genomics, 2009. **10**: p. 371.

37. Mosiolek, M., et al., *Rumex acetosa Y chromosomes: constitutive or facultative heterochromatin?* Folia Histochem Cytobiol, 2005. **43**(3): p. 161-7.

38. Siroky, J., M.R. Castiglione, and B. Vyskot, *DNA methylation patterns of Melandrium album chromosomes.* Chromosome Res, 1998. **6**(6): p. 441-6.

39. Guttman, D.S. and D. Charlesworth, *An X-linked gene with a degenerate Y-linked homologue in a dioecious plant.* Nature, 1998. **393**(6682): p. 263-6.

40. Bergero, R., et al., *Evolutionary strata on the X chromosomes of the dioecious plant Silene latifolia: evidence from new sex-linked genes.* Genetics, 2007. **175**(4): p. 1945-54.

41. Nicolas, M., et al., *A gradual process of recombination restriction in the evolutionary history of the sex chromosomes in dioecious plants.* PLoS Biol, 2005. **3**(1): p. e4.

42. Haag-Liautard, C., et al., *Direct estimation of per nucleotide and genomic deleterious mutation rates in Drosophila.* Nature, 2007. **445**(7123): p. 82-5.

43. Tamura, K., S. Subramanian, and S. Kumar, *Temporal patterns of fruit fly (Drosophila) evolution revealed by molecular clock.* Mol. Biol. Evol., 2004. **21**: p. 36-41.

44.   Bromham, L. and D. Penny, *The modern molecular clock.* Nat Rev Genet,
        2003. **4**(3): p. 216-24.
45.   Smith, S.A. and M.J. Donoghue, *Rates of molecular evolution are linked to
        life history in flowering plants.* Science, 2008. **322**(5898): p. 86-9.
46.   Charlesworth, D., *Plant sex determination and sex chromosomes.* Heredity,
        2002. **88**(2): p. 94-101.
47.   Zhang, W., et al., *DNA methylation and heterochromatinization in the male-
        specific region of the primitive Y chromosome of papaya.* Genome Res, 2008.
        18(12): p. 1938-43.
48.   Filatov, D.A. and D. Charlesworth, *Substitution rates in the X- and Y-linked
        genes of the plants, Silene latifolia and S. dioica.* Mol Biol Evol, 2002.
        19(6): p. 898-907.
49.   Perrin, N., *Sex reversal: a fountain of youth for sex chromosomes?*
        Evolution, 2009. **63**(12): p. 3043-9.
50.   Ezaz, T., et al., *Sex chromosome evolution in lizards: independent origins
        and rapid transitions.* Cytogenet Genome Res, 2009. **127**(2-4): p. 249-60.
51.   Newcomer, E.H., *The karyotype and possible sex chromosomes of Ginkgo
        biloba .* Am. J. Bot., 1954. **41**: p. 542-545.
52.   Lee, C.L., *Sex chromosomes in Ginkgo biloba.* Am. J. Bot., 1954. **41**: p. 545-
        549.
53.   Pollock, E.G., *The sex chromosomes of maidenhair tree.* J. Heredity, 1957.
        48: p. 290-294.
54.   Graves, J.A., *Sex chromosome specialization and degeneration in mammals.*
        Cell, 2006. **124**(5): p. 901-14.
55.   Warren, W.C., et al., *Genome analysis of the platypus reveals unique
        signatures of evolution.* Nature, 2008. **453**(7192): p. 175-83.
56.   Waters, P.D. and J.A. Marshall Graves, *Monotreme sex chromosomes--
        implications for the evolution of amniote sex chromosomes.* Reprod Fertil
        Dev, 2009. **21**(8): p. 943-51.
57.   Wimmer, R., et al., *Evolutionary breakpoint analysis on Y chromosomes of
        higher primates provides insight into human Y evolution. .* Cytogenet.
        Genome Res., 2005. **108**: p. 204-210.
58.   Bergero, R., et al., *Defining regions and rearrangements of the Silene
        latifolia Y chromosome.* Genetics, 2008. **178**(4): p. 2045-53.
59.   Yang, Z., *PAML 4: phylogenetic analysis by maximum likelihood.* Mol Biol
        Evol, 2007. **24**(8): p. 1586-91.
60.   Yang, Z. and R. Nielsen, *Estimating synonymous and nonsynonymous
        substitution rates under realistic evolutionary models.* Mol. Biol. Evol.,
        2000. **17**: p. 32-43.
61.   Yamato, K.T., et al., *Gene organization of the liverwort Y chromosome
        reveals distinct sex chromosome evolution in a haploid system.* Proc Natl
        Acad Sci U S A, 2007. **104**(15): p. 6472-7.

62.   Filatov, D.A., et al., *Low variability in a Y-linked plant gene and its implications for Y-chromosome evolution.* Nature, 2000. **404**(6776): p. 388-90.

63.   Filatov, D.A., et al., *DNA diversity in sex-linked and autosomal genes of the plant species Silene latifolia and Silene dioica.* Mol Biol Evol, 2001. **18**(8): p. 1442-54.

64.   Laporte, V., et al., *Indirect evidence from DNA sequence diversity for genetic degeneration of the Y-chromosome in dioecious species of the plant Silene: the SlY4/SlX4 and DD44-X/DD44-Y gene pairs.* J Evol Biol, 2005. **18**(2): p. 337-47.

65.   Filatov, D.A., *Substitution rates in a new Silene latifolia sex-linked gene, SlssX/Y.* Mol Biol Evol, 2005. **22**(3): p. 402-8.

66.   Marais, G.A., et al., *Evidence for degeneration of the Y chromosome in the dioecious plant Silene latifolia.* Curr Biol, 2008. **18**(7): p. 545-9.

67.   Kejnovsky, E., et al., *Localization of male-specifically expressed MROS genes of Silene latifolia by PCR on flow-sorted sex chromosomes and autosomes.* Genetics, 2001. **158**(3): p. 1269-77.

68.   Bachtrog, D., et al., *Genomic degradation of a young Y chromosome in Drosophila miranda.* Genome Biol, 2008. **9**(2): p. R30.

In: New Insights on Plant Sex Chromosomes   ISBN: 978-1-61470-236-8
Editor: Rafael Navajas-Pérez   ©2012 Nova Science Publishers, Inc.

Chapter IV

Papaya Sex Chromosomes

Andrea R. Gschwend and Ray Ming[*]

Department of Plant Biology, University of Illinois at Urbana-Champaign,
Urbana, IL, US

ABSTRACT

Papaya is a semi-woody tree that produces fruit rich in vitamins and minerals. It is trioecious with male, female, and hermaphrodite plants. Though many theories have been suggested in the past, papaya sex is determined by a pair of nascent sex chromosomes; females have two X chromosomes, males have an X and a Y, and hermaphrodites have an X and a Y^h, which varies slightly from the male Y. Any combination of the Y and Y^h genotype, YY, Y Y^h, or Y^h Y^h, is lethal. The X and Y chromosomes have a small non-recombining region in the centromeric and pericentromeric region. The hermaphrodite-specific region of the Y^h chromosome (HSY) is gene poor and has an increased amount of retroelements and chromosomal rearrangements compared to its X counterpart and the genome wide average. The X and Y chromosomes were estimated to have diverged relatively recently about 2-3 million years ago (MYA), and even more so for the Y and Y^h (73,000 years ago). Physical maps of the HSY and the corresponding X region have been produced and sequenced, showing the HSY sequence has expanded. Genes in these regions are being mined with a special focus on identifying the two sex determination genes, one promoting maleness and one

[*] Corresponding author: Ray Ming, E mail: rming@life.illinois.edu

suppressing femaleness. The impact of identifying the sex determination genes would be high for both the commercial production of papaya and the field of sex chromosome evolution.

INTRODUCTION

Sexual reproduction is a common phenomenon in eukaryotes. Across sexually reproducing flowering plants, a variety of breeding systems have arisen. Flowering plant species can be hermaphrodite, containing both male and female parts within the same flower, monoecious, having separate male and female flowers on the same individual, dioecious, having male and female flowers on different individuals, or a combination of the three. Hermaphroditism is the most common sexual type seen in angiosperms, whereas dioecy is only seen in about 6% of species across about 38% -of flowering plants families [1]. Dioecious angiosperm individuals were thought to have evolved independently in different families from hermaphrodite ancestors [2]. There are many possible reasons for this shift from hermaphrodite to monoecious or dioecious sex types. Hermaphrodite plants can self-pollinate, which allows for a greater opportunity to reproduce, but they do not obtain the same genetic diversity monoecious or dioecious plants acheive by cross pollinating, which gives them more evolutionary flexibility.

In dioecious plants, sex is determined at various stages of flower development, and molecularly, there are likely many different genes involved in sex determination of flowering plants. One common thread seen in many dioecious species is the presence of sex chromosomes. Sex chromosomes are those containing sex determining genes. These sex chromosomes can either be heteromorphic or homomorphic, depending on their evolutionary stage. The first report of heteromorphic sex chromosomes in plants was seen in the liverwort *Sphaerocarpos donnellii* [3]. Since then, molecular investigations have been carried out on a diversity of plants in various stages of sex chromosome evolution.

Over the years, a handful of model species for molecular studies of sex chromosomes have emerged. *Carica papaya* has been found to have nascent sex chromosomes [4]. The X and Y chromosomes are cytologically homomorphic, but the Y chromosome has a recombination suppressed male-specific region (MSY) which has lost many genes, resulting in the lethal YY genotype. Besides papaya, the evolution of the sex chromosomes of angiosperms such as strawberry, asparagus, poplar, Silene, and Rumex are also being intensely studied. Strawberry (*Fragaria virginiana*) was discovered to have two sex determination genes, one for male sterility and one for female fertility. These genes have been mapped to the same linkage group and recombination still occurs in the 5.6 cM region between the genes [5]. Asparagus (*Asparagus officinalis*) sex chromosomes are homomorphic and contain two tightly linked sex determination genes, a male activator and a female suppressor, at a single locus. Recombination in this area is suppressed, but degeneration in this region is minor, because the male YY genotype is viable [6, 7]. *Silene latifolia* has both a large male-specific non-recombining region on the Y chromosome, and pseudoautosomal regions that do still recombine with the X [8, 9]. There are at least three regions in the male-specific region of the Y relevant to sex expression, one that suppresses female-ness and two that promote maleness [10 - 13]. The sex chromosomes are heteromorphic, because the Y chromosome is larger than the X [14]. *Rumex acetosa* has a XX/XY1Y2 sex chromosome system, and sex is determined by the ratio of X-linked genes to autosomal genes [15]. As shown in these examples, there are significant differences in the structures of the sex chromosomes across these dioecious (and trioecious) species, leading to the hypothesis that these sex chromosomes vary in evolutionary stages.

Six stages of sex chromosome evolution have been proposed to explain the variation of the sex chromosomes in angiosperms [16]. In the first stage, a male-sterile mutation and a female-sterile mutation arise in close proximity on a chromosome, but recombination still occurs in this region. Strawberry is in the earliest stage of sex chromosome evolution. In the second stage, recombination is suppressed at and around the mutations, a crucial step in sex chromosome evolution, leading to degeneration of the Y chromosome, though the YY genotype is still viable, as seen in asparagus. In the third stage, the suppression of recombination spreads to other loci, forming a male-specific region on the Y chromosome, and though the chromosomes appear to be homomorphic at this stage, some genes on the Y chromosome are lost through transposable element insertions, deleterious mutations, and chromosomal rearrangements causing an inviable YY genotype. Papaya is an example of stage three. During the fourth stage, the MSY accumulates transposable elements and duplications, causing a DNA expansion. The non-recombining region spreads to the majority of the Y chromosome and the sex chromosomes are heteromorphic, with the Y chromosome often becoming considerably larger than the X chromosome. Silene is a good example of an angiosperm in stage 4. During stage five, though a small portion of the sex chromosomes continues to recombine, keeping the pair together, severe degeneration of the Y chromosome occurs and many genes lose function, leading to the loss of the non-functional sequences, causing the Y chromosome to shrink. There are no current examples of angiosperm sex chromosomes in this stage. Finally, in stage six, the suppression of recombination spreads to the entire Y chromosome, causing the Y chromosome to be completely lost. Sex is then determined by an X to autosome ratio, as is seen in Rumex.

Besides being a model species for sex chromosome studies, papaya is an important fruit crop and is widely cultivated in tropical and subtropical areas. Papaya hasa soft-wooded trunk that hollows between the nodes and is unbranched with palmate leaves [17]. It is fast growing with a short juvenile phase of about 3-8 months and can produce ripe fruit in 9 months. Papaya is also grown for the proteolytic enzyme papain, which is used as a meat tenderizer and inchill proofing beer.

Papaya is trioecious, consisting of all 3 sex types. Papaya flowers are composed of 5 waxy, ivory-white petals, which are slightly fragrant [18]. Male trees have long peduncles and inflorescences consisting of multiple slender staminate flowers. Female trees have short inflorescences with only a few flowers with large carpels and no stamens. Hermaphrodite trees also have short inflorescences, but with flowers that contain both functional carpels and stamens.

Commercially, determining the sex of the papaya trees is essential. In its vegetative form, the three sex types are indistinguishable. It isn't until the tree reaches reproductive maturity and flowers, about 4-6 months after germination, that the sex of the tree is revealed [19]. Hermaphrodite trees are favored in commercial papaya farming, because of their high productivity and because the fruit is prefered by customers. Using female trees in the field would require male trees for pollination, occupying 6-10% of valuable field space with trees that will not produce fruit. Since there are no true-breeding hermaphrodite varieties, farmers must plant 5 seedlings per plot, allow them to grow until they flower, then cut down and remove all but one hermaphrodite tree. This method is inefficient and costly. Papaya trees compete for resources, which delays production, and farmers' time and labor are wasted on plants that will ultimately be cut down. The identification of the papaya sex determination genes could lead to engineering a true breeding hermaphrodite variety to improve papaya production.

In this chapter, the past research on papaya sex chromosomes will be reviewed and the current status of the project and the implications for future research will be summarized.

MENDELIAN GENETICS OF SEX DETERMINATION

Papaya sex determination has intrigued geneticists and breeders for many years, because of papaya's unusual sex segregation ratios and because sex type is important to the commercial production of papaya. To explore sex determination in papaya, Hofmeyr in South Africa and Storey in Hawaii independently carried out a series of crosses between varying papaya sex types [20, 21]. The results were consistent among the two studies. When a hermaphrodite was self-pollinated, which is most common, the resulting offspring showed 2:1 hermaphrodite to female segregation. If a female tree was pollinated by a hermaphrodite or a male tree was crossed with a female tree, the resulting segregation was 1:1, female to hermaphrodite for the former and male to female for the latter. A hermaphrodite crossed with a male resulted in seeds segregating in a 1:1:1 ratio of males, hermaphrodites, and females. Occasionally a male flower with functional carpels would self and would result in a 2:1 male to female ratio. From these results, Storey and Hofmeyr each concluded that papaya sex is controlled by one gene with three alleles, Storey using M, M^h, and m to denote the alleles, and Hofmeyr using M_1, M_2, and m [20, 21, 22]. Since Storey's annotation differentiates between male and hermaphrodite alleles, it is more commonly used. In their hypotheses, males (Mm) and hermaphrodites (M^hm) are heterozygous, whereas the females are homozygous recessive (mm). Dominant allele combinations (MM, MM^h, M^hM^h) are lethal, causing the interesting field cross ratios observed.

These initial hypotheses sparked other explanations for sex determination in papaya. Hofmeyr expanded on his original hypothesis and suggested the chromosomes carrying the sex determination gene are sex chromosomes. He suggested the female sex determination factors primarily made up the sex chromosomes and the autosomes contain the male sex determination factors. M and M^h were thought to have an inactivated region, which was missing vital genes, causing any M and M^h combination to be lethal. Sex was thought to be determined by the balance of sex chromosomes and autosomes [23, 24].

Storey also modified his hypothesis when he took into account the long peduncles only found in males and the lethality of the homozygous dominant genotype in males and hermaphrodites, and suggested that sex was determined by a group of closely linked genes which are clustered on a small region of the sex chromosome and behave as a unit, because recombination does not occur in this section [25]. A handful of genes or factors were hypothesized to be present in this region. *Mp*, the male peduncle gene, was thought to be a gene that causes the long peduncles of male flowers. The zygotic lethal factor, *l*, causes the lethal phenotype seen in males and hermaphrodites. The androecium is suppressed when *sa* is homozygous recessive, like it is in females. The suppressor of the gynoecium, *sg*, controls the suppression of carpel development when present as homozygous recessive. Because of the sex reversals seen in the field, it was suggested that males and hermaphrodites are only heterozygous for this gene. *C*, the hypothetical suppressor of recombination, is a factor that prevents recombination in this region. Given this information, the genotypes of *M*, M^h, and *m* would be as follows, though the order of these genes is arbitrary:

$$Mm = \frac{Mp \; l \; C \; + \; sg}{+ \; + \; + \; sa \; +}$$

$$M^h m = \frac{+ \; l \; C \; + \; sg}{+ \; + \; + \; sa \; +}$$

$$mm = \frac{+ \; + \; + \; sa \; +}{+ \; + \; + \; sa \; +}$$

A few years later it was suggested by Horovitz and Jimenez (1967) that the papaya sex determination is an XX-XY system, where the female genotype was XX, the male was XY and the hermaphrodite was XY_2, with a slightly different Y chromosome. The Y chromosome contained a lethal factor causing YY, YY_2, and Y_2Y_2 genotypes to abort. These conclusions were drawn from intergeneric hybridizations done between *Carica* and *Vasconcellea* species [26]. Around the same time, Hamilton and Izuno identified a female tree with long peduncles, with the majority of flowers being female, but containing a few male and hermaphrodite flowers. Because of reciprocal cross data from resulting progeny, they hypothesized that there are only two sexes, females and a sex type that is variable, and can range from hermaphrodite to male with a number of intermediates [27]. This discovery caused Storey to revise his model by removing *Mp* and changing *sa* and *sg* [28]. *SA* now converts the ancestral androecium to the current gynoecium, *sa* causes normal androecium development, *SG* controls male carpel abortion, and *sg* allows for the expression of SA, controlling normal flower abortion. The lethal factor (*l*) and suppressor of recombination (*C*) are the same.

In this model, the female was considered homozygous for SA and sg, and the variable sex had a copy of SA, sa, SG, and sg.

More recently, Sondur et al. (1996) suggested papaya sex determination is controlled by trans-regulatory elements. The dominant male allele (*SEX1-M*) encodes a transacting factor that promotes stamen development, but suppresses carpel development. The dominant hermaphrodite allele (*SEX1-H*) also promotes stamen development, but reduces carpel size. The recessive female allele (*sex1*-f) is unable to promote stamens. The lethal factor was attributed to the lack of a vital function that is present *in sex1-f*, but not in *SEX1-M* and *SEX1-H* [29].

These hypotheses were stepping stones leading to the current research being done in exploring the sex of papaya. When the proper technology was developed, molecular studies began to shed light on the mysteries surrounding papaya sex and eventually led to the discovery that papaya does in fact have incipient sex chromosomes, with a sex-specific non-recombining region. The next section will follow the molecular experiments leading to this discovery, as well as the current knowledge about papaya sex chromosomes.

MOLECULAR GENETICS OF SEX CHROMOSOMES

Though many hypotheses had been made about the sex determination system of papaya, little concrete molecular data had been generated to verify which hypothesis accurately described what was occurring in papaya that led to these three sex types. Genetic cross data, phenotypic data, and some early cytological observations were the only evidence used to form these early hypotheses. It wasn't until the applications of molecular techniques and biotechnology, that scientists had the means to tackle the question of papaya sex determination.

The first method explored to detect papaya sex was the use of sex-linked molecular markers. Microsatellite and sequence-characterized amplified region (SCAR) markers, which showed different banding patterns between the sex types, were successfully developed by different papaya research groups. This allows papaya sex to be determined during the vegetative state, but for commercial use, it is too costly to test thousands of seedlings and relocate them to the field [30 - 33].

Molecular markers were also used in producing multiple genetic linkage maps for papaya. The first map was constructed by Hofmeyr (1939), consisting of three morphological markers: sex, flower color, and stem color [23]. The second genetic linkage map consisted of 62 randomly amplified polymorphic DNA

(RAPD) markers and mapped sex onto linkage group 1 [29]. The third map encorporated 1,498 amplified fragment length polymorphism (AFLP) markers, the papaya ringspot virus coat protein marker, sex, and fruit flesh color, totaling 1,501 markers which were mapped onto 12 linkage groups [34]. Most recently a high density genetic map using 712 simple sequence repeat (SSR) markers, designed from BAC end sequences and whole-genome shot gun sequences, and a morphological marker resulted in 9 large linkage groups and 3 small linkage groups [35]. Sex was mapped onto linkage group 1, one of the 9 large linkage groups.

The construction of the papaya hermaphrodite BAC library, made up of 39,168 clones with an average insert size of 132 kb and 13.7X genome coverage, allowed for a new depth of exploration of the papaya sex chromosomes [36]. The HSY was mapped with 225 sex co-segregating AFLP markers on linkage group 1, showing severe suppressed recombination in this sex determining region. SCAR markers were developed from sex co-segregating AFLP markers, and were used to screen the BAC library for physical mapping. Those BACs were extended by designing probes from their BAC end sequences to scan the BAC library for overlapping BACs. Through chromosome walking methods, a rudimentary physical map was produced with 2 major and 3 minor contigs that spanned 2.5Mb [4]. These efforts resulted in the discovery that severe suppression of recombination and degeneration is occurring in 10% of these homologous chromosomes, leading to the conclusion that these are in fact incipient sex chromosomes [4]. Next, 50,661 BAC ends from 26,017 BAC clones were sequenced, allowing for chromosome walking techniques to be implicated to identify additional BACs in this area of interest [37].

The completion of the papaya whole genome sequence provided the resources to expedite the physical map construction for the hermaphrodite HSY and X-specific regions [38]. The hermaphrodite BAC library clones were fingerprinted and BACs associated with the 2.5Mb physical map were used to discover contigs in the whole genome physical map that could aid in expanding the HSY through chromosome walking. Probes were designed from HSY BACS to detect corresponding X and male Y BACs to produce maleMSY and corresponding X specific physical maps, as well.

Some of the BACs located on the HSY, as well as a selection of paired X- and Y- specific BACS, were sequenced and investigated. The HSY BACs showed a deficiency of genes, a large number of retroelements, and gene duplication events compared to the X [39]. Using genes found on both the HSY and X BACs, the divergence time of the X and Y^h chromosomes was estimated to be between 0.5 and 2.5 MYA, suggesting the sex chromosomes evolved at the genus or species level [40]. The Y and Y^h sequences were found to be nearly identical and likely arose from the same ancestral chromosome, instead of evolving separately. The divergence time between the Y and Y^h chromosomes was predicted to be 73,000 years [41]. In the male specific regions of the compared BACs, various chromosomal rearrangements have occurred, such as inversions, deletions, insertions, duplications and translocations

To date, the HSY and corresponding X region of the hermaphrodite have been physically mapped. Each physical map has only one remaining gap. The HSY physical map has a gap along Border A that has been filled on the X physical map. The corresponding X region has a gap located towards the center of the physical map between BACs 136D11 and 08K16, which is filled on the HSY physical map. The HSY spans ~8Mb and the X spans ~5Mb.

CYTOGENETICS OF SEX CHROMOSOMES

Of the nine chromosome pairs of papaya, seven are metacentirc and the remaining two pairs are submetacentric [17]. Papaya chromosomes are small and uniform in morphology, making them hard to differentiate using length, arm ratios, or banding patterns [42]. They are also cytologically homomorphic and were difficult to identify in early studies of papaya sex determination. In an early investigation, precocious separation was observed between a chromosome pair during anaphase I of meiosis of a pollen mother cell in males and hermaphrodites [43]. Recombination occurs throughout the homologous regions of the sex chromosomes, but it is suppressed in the sex specific region of the chromosome pairs (~13% of the sex chromosomes) [44]. Recombination rate recovers and elevates in the border regions of the HSY [45].

To link chromosomes to their genetic sequences, genetic mapping of the papaya genome was carried out and resulted in 12 genetic linkage groups, including nine major and three minor linkage groups [35]. The three minor linkage groups 10, 11, and 12, were merged with major linkage groups 8, 9, and 7, respectively, using molecular cytogenetic approaches [42]. To determine which chromosome corresponds to which genetic linkage group, linkage group-specific BACs were used as probes for fluorescence in situ hybridization (FISH) in papaya meiotic pachytene chromosomes [46]. The X and Y chromosomes were identified as the second longest chromosome pair and were designated as chromosome 1 in the karyotype. The remaining papaya chromosomes were numbered according to length, chromosome 2 being the longest and chromosome 9 being the shortest.

To locate the HSY region on the sex chromosomes, 2 confirmed BACs in this area were directly hybridized to interphase, prometaphase, metaphase, and anaphase chromosomes [39]. These BACs hybridized near the centromere of the Y^h. One BAC had a weaker signal on the X, which suggested the sequences on the HSY and X in this region were still relatively conserved. Since the second BAC only hybridized to the Y^h, likely that BAC sequence had diverged considerably. Pachytene FISH was also utilized to map one of these HSY-specific BACs, along with its neighboring BACs, and a non-HSY BAC. The HSY BACs showed strong signals only on the $Y^{h,}$ whereas the non-HSY BAC showed a strong signal on a different homologous pair of chromosomes. This study verified the identity of the X and Y^h chromosomes and located the HSY near the centromere on the Y^h.

To further explore the structure of the MSY and X regions, hermaphrodite meiotic pachytene chromosomes were stained with 4',6-diamidino-2-phenylindole (DAPI), which stains heterochromatic regions of chromosomes [44]. Based on the staining, the arms of the chromosomes were mostly euchromatic, but clusters of heterochromatin were found around the centromere. Specifically, the XY^h bivalent was mostly euchromatic, with the X chromosome being the most euchromatic chromosome in the papaya genome, but five knob-like regions of heterochromatin, numbered K1 though K5, were found in the HSY^h [44, 46]. The largest knob, K1, was shared between the HSY^h and the X, but K2-K5 were only found on the HSY. The knobs were also found to be highly methylated. The heterochromatic knobs were likely the result of transposable elements and the high DNA methylation in these regions could be a defense mechanism against transposable element invasion. The HSY contained two small regions of 5S rDNA, which is an element of the large subunit of the ribosome involved in translation [46]. These regions were associated with K2 and K4. The X chromosome did not exhibit 5S rDNA. The accumulation of 5S rDNA in the HSY likely led to the materialization of heterochromatin and assisted in the differentiation of the sex chromosomes [46].

During X and Y^h chromosomal pairing, a slight curve in the Y^h chromosome occurred to allow for pairing [44]. The region around K4 had accumulated considerably more DNA then it's X corresponding region, causing the curving of the Y^h chromosome during pairing. By implementing meiotic metaphase I-based

FISH using Y^h-specific BACs, the centromere of the Y^h chromosome was found to be located in the HSY, specifically associating the centromere with K4 [44]. The area around K4 showed more sequence divergence from the X than other regions of the HSY.

CONCLUSIONS

The discovery of incipient sex chromosomes in papaya offered opportunities to investigate the early events of sex chromosome evolution. With the completion of the HSY and corresponding X physical maps, the genes present in these regions can be further explored. The most coveted genes to discover are the sex determination genes. The current hypothesis is that there are two key genes involved in determining sex in papaya. The first is a stamen promoting gene, which would be found on the MSY of the male Y and the HSY of the hermaphrodite Y^h. The other gene involved in sex determination is a gain of function carpel suppressing gene, which would only be located on the MSY of the male Y. A deletion mutant that displayed a male to female sex reversal phenotype has allowed for the location of the two sex determination genes to be narrowed down to a 1Mb region on the MSY (R. Ming and Q. Yu, unpublished data). This region has been thoroughly mined for genes and candidate genes have already been identified. The identification of these two genes will eventually lead to the development of a true breeding hermaphrodite papaya variety, solving the problems of planting multiple seedlings in the field.

Further exploration of the papaya sex chromosomes will also provide insight into early stages of sex chromosome evolution. Papaya is in the Caricaceae family, consisting of 35 species, including 32 dioecious, 2 trioecious, and 1 monoecious species. The sequence data accumulated for papaya will allow for genic comparisons across the species in Caricaceae. The majority of the Caricaceae family is dioecios, suggesting that dioecy arose in this family after the divergence from its closest family Maringaceae about 60 MYA. By comparing the genes and genomic sequences of the homologous chromosomes within the family Caricaceae, a better understanding of how sex chromosomes evolve over time, as well as what features were conserved and lost, and what features were present before the divergence of the genera and species. Since the Y and Y^h sequences of papaya were estimated to diverge a mere 73,000 years ago, it would be interesting to investigate the causes of this reversion back to hermaphrodite.

The knowledge gained from the sex chromosomes in papaya can be applied, in general, to how sex chromosomes emerged and evolved. By studying the sequences of the papaya HSY and X regions, the initial cause of suppression of recombination, as well as the genes determining sex, can be discovered. The characteristics of papaya sex chromosomes can be compared with other sex chromosome model species in varying evolutionary stages, such as strawberry and Silene, to uncover the mechanisms and driving forces of sex chromosome evolution.

REFERENCES

[1]   Renner SS, Ricklefs RE (1995) Dioecy and its correlates in the flowering plants. *Am J Bot* 82:596-606.
[2]   Takhtajan A (1969) Flowering plants: origin and disposal. Oliver and Boyd, Edinburgh, UK.
[3]   Allen CE (1917) A chromosome difference correlated with sex differences in *Sphaerocarpos*. *Science* 46:466-7.
[4]   Liu Z, Moore PH, Ma H, Ackerman CM, Ragiba M et al. (2004) A primitive Y chromosome in papaya marks incipient sex chromosome evolution. *Nature* 427:348-52.
[5]   Spigler RB, Lewers KS, Main DS, Ashman TL (2008) Genetic mapping of sex determination in a wild strawberry, *Fragaria virginiana*, reveals earliest form of sex chromosome. *Heredity.* 101:507-17.
[6]   Marks M (1973) A reconsideration of the genetic mechanism for sex determination in *Asparagus officinalis. Proceedings of the Eucarpia meeting on asparagus (Asparagus officinalis L.).* Versailles, pp 123-28.
[7]   Reamon-Büttner SM, Schondelmaier J, Jung C (1998) AFLP markers tightly linked to the sex locus in *Asparagus officinalis* L. *Mol. Breed.* 4:91-8.
[8]   Scotti I, Delph LF (2006) Selective trade-offs and sex-chromosome evolution in *Silene Latifolia. Evolution* 60:1793-1800.
[9]   Delph LF, Arntz AM, Scotti-Saintagne C, Scotti I (2010) The genomic architecture of sexual dimorphism in the dioecious plant Silene latifolia. *Evolution* 64:2873-86.
[10] Donnison IS, Siroky J, Vyskot B, Saedler H, Grant SR (1996) Isolation of Y chromosome-specific sequences from Silene latifolia and mapping of

male sex-determining genes using representational difference analysis. *Genetics* 144:1893-901.

[11] Lardon A, Georgiev S, Aghmir A, Le Merrer G, Negrutiu I (1999) Sexual dimorphism in white campion: complex control of carpel number is revealed by Y chromosome deletions. *Genetics* 151:1173-85.

[12] Lebel-Hardenack S, Hauser E, Law TF, Schmid J, Grant SR (2002) Mapping of sex determination loci on the white campion (*Silene latifolia*) Y chromosome using amplified fragment length polymorphism. *Genetics* 160:717-25.

[13] Zluvova J, Georgiev S, Janousek B, Charlesworth D, Vyskot B, Negrutiu I (2007) Early events in the evolution of the *Silene latifolia* Y chromosome: male specialization and recombination arrest. *Genetics* 177:375-86.

[14] Grabowska-Joachimiak A, Joachimiak A (2002) C-banded karyotypes of two Silene species with heteromorphic sex chromosomes. *Genome* 252:243-52.

[15] Lengerova M, Vyskot B (2001) Sex chromatin and nucleolar analyses in *Rumex acetosa* L. *Protoplasma* 217:147-53.

[16] Ming R, Bendahmane A, Renner SS (2011) Sex chromosomes in land plants. *Annu Rev Plant Biol* 62:485-514.

[17] Ming R, Yu Q, Blas A, Chen C, Na JK, Moore PH (2008a) Plant genetics and genomics: crop and models, Chapter 16: Genomics of papaya, a common source of vitamins in the tropics. *Genomics of tropical crop plants*. Springer, New York, pp 405-20.

[18] Morton, J (1987) Papaya. In: *Fruits of warm climates*. Creative Resource Systems, Inc., Miami, FL, pp 336-46.

[19] Ming R, Yu Q, Moore PH (2007) Sex determination in papaya. *Semin Cell Dev Biol* 18:401-8.

[20] Hofmeyr JDJ (1938) Genetical studies of *Carica papaya* L. I. The inheritance and relation of sex and certain plant characteristics. II. Sex reversal and sex forms. *S Afr Dept Agri and Sci Bul* 187:64.

[21] Storey WB (1938) Segregation of sex types in Solo papaya and their application to the selection of seed. *Proc Am Soc Hort Sci* 35:83-5.

[22] Storey WB (1941) The botany and sex relations of the papaya. *Hawaii Agr Exp Sta Bul* 87:5-22.

[23] Hofmeyr JDJ (1939) Sex reversal in *Carica papaya* L. *S Afr J Sci* 26:286-7.

[24]   Hofmeyr JDJ (1967) Some genetic breeding aspects of *Carica papaya* L. *Agron Trop.* 17:345-51.

[25]   Storey WB (1953) Genetics of papaya. *J Hered* 44:70-8.

[26]   Horovitz S, Jiménez H (1967) Cruzamientos interspecificos e intergenericos en caricaceas y sus implicaciones fitotechicas. *Agron Trop* 17:323-43.

[27]   Hamilton RA, Izuno T (1967) A revised concept of sex inheritance in *Carica papaya*. *Agron Trop* 17:401-2.

[28]   Storey WB (1976) Papaya. In: Simmonds NW (eds) *Evolution of crop plants*. Longman, London and New York, pp 21-24.

[29]   Sondur SN, Manshardt RM, Stiles JI (1996) A genetic linkage map of papaya based on randomly amplified polymorphic DNA markers. *Theor Appl Genet* 93:547-53.

[30]   Parasnis AS, Ramakrishna W, Chowdari KV, Gupta VS, Ranjekar PK (1999) Microsatellite (GATA)n reveals sex-specific differences in Papaya. *Theor Appl Genet* 99:1047-52.

[31]   Parasnis AS, Gupta VS, Tamhankar SA, Ranjekar PK (2000) A highly reliable sex diagnostic PCR assay for mass screening of papaya seedlings. *Mol Breed* 6:337-44.

[32]   Urasaki N, Tokumoto M, Tarora K, Ban Y, Kayano T et al. (2002) A male and hermaphrodite specific RAPD marker for papaya (*Carica papaya* L.) *Theor Appl Genet* 104:281-5.

[33]   Deputy JC, Ming R, Ma H, Liu Z, Fitch MMM et al. (2002) Molecular markers for sex determination in papaya (*Carica papaya* L.) *Theor Appl Genet* 106:107-11.

[34]   Ma H, Moore PH, Liu Z, Kim MS, Yu Q et al. (2004) High-density linkage mapping revealed suppression of recombination at the sex determination locus in papaya. *Genetics* 166:419-36.

[35]   Chen C, Yu Q, Hou S, Li Y, Eustice M et al. (2007) Construction of a sequenced-tagged high-density genetic map of papaya for comparative structural and evolutionary genomics in Brassicales. *Genetics* 177:2481-91.

[36]   Ming R, Moore PH, Zee F, Abbey CA, Ma H, Paterson AH (2001) Construction and characterization of a papaya BAC library as a foundation for molecular dissection of a tree-fruit genome. *Theor Appl Genet* 102:892-9.

[37]   Lai CW, Yu Q, Hou S, Skelton RL, Jones MR et al. (2006) Analysis of papaya BAC end sequences reveals first insights into the organization of a fruit tree genome. *Mol Genet Genomics* 276:1-12.

[38] Ming R, Hou S, Feng Y, Yu Q, Dionne-Laporte A et al. (2008b) The draft genome of the transgenic tropical fruit tree papaya (Carica papaya Linnaeus). *Nature* 452:991-6.

[39] Yu Q, Hou S, Hobza R, Feltus FA, Wang X et al. (2007) Chromosomal location and gene paucity of the male specific region on papaya Y chromosome. *Mol Genet Genomics* 278:177-85.

[40] Yu Q, Hou S, Feltus A, Jones MR, Murray JE et al. (2008a) Low X/Y divergence in four pairs of papaya sex-linked genes. *Plant J* 53:124-32.

[41] Yu Q, Navajas-Perez R, Tong E, Robertson J, Moore PH, et al. (2008b) Recent origin of dioecious and gynodioecious Y chromosomes in papaya. *Trop Plant Biol* 1:49-57.

[42] Wai CM, Ming R, Moore PH, Paull RE, Yu Q (2010) Development of chromosome-specific cytogenetic markers and merging of linkage fragments in papaya. *Trop Plant Biol* 3:171-81.

[43] Kumar LSS, Abraham A, Srinivasan VK (1945) The cytology of *Carica papaya* Linn. *Indian J Agr Sci* 15:242-53.

[44] Zhang W, Wang X, Yu Q, Ming R, Jiang J (2008) DNA methylation and heterochromatinization in the male-specific region of the primitive Y chromosome of papaya. *Genome Res* 18:1938-43.

[45] Yu Q, Tong E, Skelton RL, Bowers JE, Jones MJ, et al. (2009) A physical map of the papaya genome with integrated genetic map and genome sequence. *BMC Genomics* 10:371.

[46] Zhang W, Wai CM, Ming R, Yu Q, Jiang J (2010) Integration of genetic and cytological maps and development of a pachytene chromosome-based karyotype in papaya. *Trop Plant Biol* 3:166-70.

In: New Insights on Plant Sex Chromosomes ISBN: 978-1-61470-236-8
Editor: Rafael Navajas-Pérez ©2012 Nova Science Publishers, Inc.

Chapter V

Fragaria: A Polyploid Lineage for Understanding Sex Chromosome Evolution

Tia-Lynn Ashman[*]*, Rachel B. Spigler, Margot T. Goldberg,*
and Rajanikanth Govindarajulu
Department of Biological Sciences, University of Pittsburgh, US

ABSTRACT

Fragaria is an exceptional model system for understanding sexual system and sex chromosome evolution. The genus hosts the entire range of sexual systems, as well as novel features, such as diversity of ploidy levels and the presence of female heterogametey. Comparative genetic mapping has revealed the autosomal ancestor of sex chromosomes in two species and exposed the evolutionary lability of sex-determining chromosomes in this young system. Recent QTL analyses have allowed the first glimpse of the genomic architecture of secondary sexual dimorphism and point the way for future work aimed at understanding the driving forces behind sex chromosome initiation and turnover. The genus is also poised to aid in answering the question of how genome doubling and merger facilitates the initiation of polymorphic sexual systems and the development of sex chromosomes.

* Author for correspondence: tia1@pitt.edu, 412-624-0984

INTRODUCTION

Completely separate sexes (males and females) have evolved repeatedly from hermaphroditic ancestors in flowering plants, and therefore certain groups of taxa can provide insight into the process of sex chromosome evolution [1]. Pairs of closely related taxa that differ in sexual system, i.e., species with combined sexes versus separate sexes, can inform on the origin of sex chromosomes from autosomes, as well as provide evidence for the initial steps in sex chromosome evolution, including linkage of and recombination suppression between sex function loci. By the same token, comparisons between two species with different degrees of sexual differentiation can allow investigation of the genomic mechanisms and consequences of recombination suppression as well as potentially expose the evolutionary lability of sex-determining chromosomes. Lastly, such comparative work can inform on the genesis of secondary sexual dimorphism and the role of sexually antagonistic selection in the evolution of sex chromosomes.

Fragaria is a genus that is poised to offer insight into all these areas as well as into the potential role that genome doubling and merger (i.e., polyploidy) may play in sexual system and sex chromosome evolution. In this chapter, we begin with a brief overview of the sexual system diversity in *Fragaria* and our current understanding of the genetics of sex expression in this genus. We then take a more in-depth look at recent genetic mapping studies which have opened the door to understanding the evolution of sex chromosomes in the genus as well as offered the first glimpse of the genetic architecture of sexual dimorphism. We conclude with some predictions for how polyploidy may contribute to sexual system evolution and fruitful avenues for further research.

SEXUAL SYSTEM DIVERSITY IN FRAGARIA

The genus *Fragaria* (Rosaceae) contains the whole range of diversity in sexual systems [2]. Two-thirds of the 27 wild species (or subspecies) are hermaphroditic and either self-compatible (10 species) or self-incompatible (7 species). Ten species show some degree of sexual polymorphism, with gynodioecy (females and hermaphrodites) in one species, subdioecy (females, hermaphrodites and males) in three species, and dioecy (females and males) in six others (Table 1).

Table 1. Species in the genus *Fragaria*, along with their sexual system, mating system, ploidy level, as well as phylogenetic and geographic distribution. Data in the table is based on several sources [2, 5, 10, 16, 18, 27, 60, 61].

Species	Sexual system[1]	Mating system[2]	Ploidy[3]	Phylogenetic membership Clade[4]	Clade[5]	Geographic range[6]
F. vesca ssp. *vesca*	H	SC	2x	A	Y	Northern hemisphere
F. vesca ssp. *braceata*	G	SC	2x	A	na[7]	Western N. America
F. vesca ssp. *americana*	H	SC	2x	na	Y	Eastern N. America
F. vesca ssp. *californica*	H	SC	2x	na	na	California, S. Oregon
F. viridis	H	GSI	2x	A	Y	Europe-Siberia
F. yezoensis	H	SC	2x	na	X	Japan
F. nipponica	H	GSI	2x	B	X	Japan
F. nubicola	H	GSI	2x	A	X	Himalayas
F. bucharica	H	GSI	2x	A	na	Western Himalayas
F. daltoniana	H	SC	2x	B	X	Himalayas
F. nilgerrensis	H	SC	2x	D	X	Southeastern Asia
F. mandshurica	H	GSI #	2x	A	Y	North China
F. pentaphylla	H	GSI	2x	B	X	North China
F. chinensis	H	GSI	2x	B	na	North China
F. iinumae	H	SC	2x	C	Z	Japan
F. xbifera	H	SC	2x, 3x	A	na	Europe
F. gracilis	D	SC	4x	B	X	North China
F. corymbosa	D	SC	4x	B	X	North China
F. moupinensis	D	SC	4x	B	X	South West China
F. orientalis	SD	SC	4x	A	Y	Northeastern Asia
F. tibetica	D	SC	4x	B	X	Eastern Himalayas
F. moschata	SD or D*	SC	6x	A	Y	Europe-Siberia
F. chiloensis	D*	SC	8x	A	Y & Z	Western N. and S. America
F. chiloensis spp. *sandwichensis*	H	SC	8x	na	na	Hawaii
F. virginiana	SD	SC	8x	A	Y & Z	North America
F.xcuneifolia	D*	SC	8x	A	na	Western N. America
F. iturupensis	H	SC	8x, 10x	A	Y & Z	Iturup Island

[1] H=hermaphroditism, D=dioecy (females and males), D*= dioecy with rare hermaphrodites, G=gynodioecy (females and hermaphrodites), SD=subdioecy (females, hermaphrodites, males).

[2] SC=self compatible, GSI = gametophytic self incompatibility. GSI# =partial GSI.

[3] 2x=diploid (x=7), 4x=tetraploid, 6x=hexaploid, 8x=octoploid, 10x=decaploid.

[4] Clade designations based on chloroplast phylogenies [12-14].

[5] Clade designations based on nuclear phylogeny [15].

[6] Based on [2].

[7] na= information not available.

Additionally, the cultivated species *F.×ananassa* is a cross between two sexually polymorphic species (*F. virginiana* and *F. chiloensis*) and has been artificially selected to be hermaphroditic.

It is worth noting, however, that the distinctions between these polymorphic sexual systems can be blurred because characterizing plant sexual phenotype (gender) is often difficult. In the case of strawberry, a qualitative characterization of plant gender from the perspective of male function can be straightforward, i.e., 'presence' versus 'absence' of pollen production in anthers, where the absence of pollen production regardless of the size of the anther would indicate male-sterility. However, characterizing female function can be more difficult because sterility can occur at different levels, e.g., at the level of flowers, which may be estimated as the proportion of flowers that yield a fruit 'fruit set' (note, that here fruit refers to the entire aggregate fruit that results from the enlarged receptacle) or at the level of ovules, which may be estimated as the proportion of ovules that yield a seed or 'seed set' (note, that here seed refers to the uniovulate achene). These estimators of female function can both vary quantitatively and be influenced by aspects of the environment, including pollination adequacy and resource availability [3-7]. As a consequence, there has been a wide range of terminology used to characterize strawberry plants that bear pollen but also produce some fruit. For example those with high and consistent fruit set have been referred to as 'strong hermaphrodites' or simply 'hermaphrodites', whereas those with low and inconsistent fruit set have been called 'weak hermaphrodites,' 'polygamodioecious', or 'inconstant males' and those that set no fruit but still produce vestigial female organs have been called 'functional males' or 'staminate' plants [3, 8, 9]. Moreover, the presence of plants that do not produce pollen and do not set fruit (i.e., neuters) can easily be mistaken for unpollinated females, and thus be overlooked. Neuters have been observed in wild populations and produced in controlled crosses in at least one species (*F. virginiana;* Ashman, pers. obs. [4, 10]) and they can be a particularly informative phenotypic class for understanding the genetic architecture of sex determination.

Not only is plant gender at the individual level notoriously difficult to characterize but classifying polymorphic sexual systems can also be problematic because transitions between sexual systems at the population or species level can be gradual rather than abrupt. For instance, because gynodioecy can evolve into subdioecy when males invade [11], the frequency of the sex morphs in populations of these species can vary dramatically-- some populations contain only hermaphrodites whereas others contain two or three sexually functional morphs in varying frequencies (e.g., *F. virginiana* [3, 5]). Moreover, even predominantly dioecious species such as *F. chiloensis,* can occasionally have populations with hermaphrodites [8]. In addition, *F. chiloensis* has one island subspecies that is described as entirely hermaphroditic. Thus, in Table 1, we have noted species as subdioecious if population surveys have documented substantial frequencies of hermaphrodites, but acknowledge that in the literature these may have been referred to as trioecious, or polygamodioecious. We have also noted cases of dioecious species where there are reports of rare hermaphrodites, or where controlled crosses have produced hermaphrodites.

Across *Fragaria*, polymorphic sexual systems are mostly found amongst the polyploid species. With the exception of *F. vesca* ssp. *bracteata* which is gynodioecious, all diploids are exclusively hermaphroditic. Two other phylogenetic patterns are also evident in the genus. First, dioecy exists in all major clades identified by chloroplast [12-14] or nuclear gene [15] phylogenies (Table 1), suggesting that it has arisen at least twice and possibly as many as four times [16]. Second, polyploid dioecious taxa are often sister to, or nested with, taxa that are self-incompatible, hermaphroditic, and diploid [13]. This latter pattern is particularly intriguing and might suggest that some aspect of genome doubling facilitates the evolution of dioecy [17].

SEX DETERMINATION IN FRAGARIA

Gaining an understanding of sex determination in *Fragaria* has engaged scientists for centuries. Early classical Mendelian genetic studies [4, 18-20] were instrumental in identifying one of the single most important features of sex expression in strawberries, that is, the characterization of the male-sterility mutation as dominant to the male-fertility allele in several species, including the one diploid gynodioecious species *F. vesca* ssp. *bracteata* (Table 2). Dominant male sterility is rare in plants [21], but it can lead to females being the heterogametic sex and ultimately a ZW sex determination system, which is also quite rare in plants [11, 22, 23].

Both dominant and recessive female sterility have been found in strawberry. In tetraploid *F. orientalis* and hexaploid *F. moschata*, the female-sterility mutation is dominant to the wild type that confers female fertility.

Table 2. Summary of the state of knowledge of the genetics of sex expression in sexually polymorphic species of *Fragaria*

Species	Ploidy	Inheritance	Male sterility	Female sterility	Chromosomal location[1]	Citation
F. vesca ssp. *bracteata*	2x	Disomic	Dominant	None	unknown	18
F. orientalis	4x	Tetrasomic	Dominant	Dominant	unknown	19
F. moschata	6x	Hexasomic	Dominant	Dominant	unknown	24
F. virginiana	8x	Disomic	Dominant	Recessive	Top VI-C	10, 18
F. chiloensis	8x	Disomic	Dominant	Recessive	Bottom VI-A	18, 27

[1]linkage group number.

Moreover, the current hypothesis for sex expression in *F. orientalis* and *F. moschata* involves tetrasomic and hexasomic inheritance, respectively, with three alleles segregating at a single locus ($Su^F=$ 'male suppressor', Su^+ 'hermphroditism', Su^M 'female supressor') [24]. By contrast, in the two octoploid species, *F. chiloensis* and *F. virginiana,* inheritance of sex expression is disomic and female sterility is recessive to female fertility (Table 2). Although a similar single locus model (with three alleles: femaleness '*F*' dominant to hermaphroditism '*h*' and maleness '*m*') was proposed by Ahmadi and Bringhurst [18] to explain sex expression in these two octoploid species, this model was re-evaluated using genetic mapping and by scoring male and female function separately. Specifically, Spigler and colleagues [10] scored individuals as 'male sterile' if they produced small vestigial stamens and anther sacs devoid of pollen, and as 'male fertile' if they produced large stamens with plump, pollen-filled anther sacs. Female function was based on the proportion of flowers that set fruit under hand outcross pollination, and to facilitate the initial mapping Spigler et al. [10] converted this quantitatively varying trait to a qualitative one. They scored plants as 'female sterile' when they set < 5% of their flowers into fruits and 'female fertile' when they set ≥ 5% of their flowers into fruits. This approach revealed that sexual function in *F. virginiana* is determined by two linked sex function 'loci', each with a sterility allele. Moreover, the presence of neuter individuals (both male and female sterile) was diagnostic of recombination between the two sex function loci in this cross. Subsequent to this original mapping effort, Spigler and colleagues [25] mapped female function as a qualitative trait and confirmed the primacy of this location in determining female function (i.e., the QTL explained 95% of the variation in fruit setting ability).

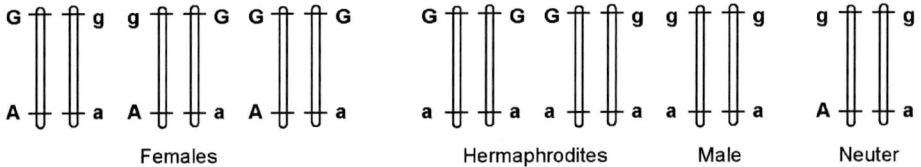

Females Hermaphrodites Male Neuter

Figure 1. Two-locus model of sex determination in *Fragaria*. Two linked but recombining sex function 'loci': Locus A represents male function: sterility 'A' is dominant to fertility 'a'; Locus G represents female function: sterility 'g' is recessive to fertility 'G'.

Accordingly, these authors proposed a two locus model for sex determination in *Fragaria* (Figure 1: one locus for male function, where sterility 'A' is dominant to fertility 'a', and one locus for female function where sterility 'g' is recessive to fertility 'G'). This model can account for the four sexual phenotypes found in subdioecious species (i.e., females [2 possible genotypes: AG|aG, AG|ag (or Ag|aG)], hermaphrodites [2 possible genotypes: aG|ag, aG|aG], males [ag|ag] and neuters [Ag|ag]), and explain the variation in sexual phenotypes observed in wild *F. virginiana* populations [5]. This model also provides a framework for understanding how recombination suppression between the two sex function loci can lead to dioecy. Specifically, once both male and female sterility exist in a population selection for restricted recombination between the two sex function genes arises because recombination in females carrying the female sterility allele (AG|ag) creates gametes that carry both sterility alleles (Ag) or both fertility alleles (aG). When these are fertilized by gametes from males (ag|ag) or hermaphrodites carrying the female sterility allele (aG|ag) they create neuters (Ag|ag) with no reproductive fitness and hermaphrodites (i.e., aG|aG) that are expected to be at a selective disadvantage under the current sexual system, i.e., subdioecy [11]. When males supplant all hermaphrodites full dioecy evolves, wherein one sex chromosome carries the male-sterility allele in coupling with the female-fertility allele (AG) that produces females and the other sex chromosome carries the female-sterility allele in coupling with the male-fertility allele (ag) that produces males [26]. As a result, populations contain only males (ag|ag) and females with the genotypic configuration AG|ag.

Further to this work, Goldberg and colleagues [27] created a genetic map for *F. chiloensis* and confirmed similar dominance relations of sterility alleles at the male and female function loci. But they also identified two important differences from *F. virginiana*. First, segregation analysis of the nearly 100 progeny from the cross did not detect recombination between the sex loci, a finding that could support either the single- or two-locus model. But perhaps more importantly, it revealed that the chromosomal location of the sex-determining region in *F. chiloensis* was different from that in *F. virginiana*, potentially indicating an evolutionary divergence between the species' sex chromosomes that involved more than just recombination suppression (see also below).

ORIGIN AND DIVERSITY OF SEX CHROMOSOMES IN FRAGARIA

Comparative genetic linkage mapping has provided clues to the origin and potential diversity of sex chromosomes in strawberry. Spigler and colleagues [28] compared their map of subdioecious *F. virginiana* to one based on a cross between two hermaphroditic diploid *Fragaria* (*F. vesca* and *F. nubicola*, hereafter "FV×FN" [29]). The extensive macrosynteny between the octoploid and diploid maps, allowed for unequivocal establishment of the sex-determining chromosome of *F. virginiana* (VI-C) as a homoeolog of autosomal chromosome 6 in the diploids (Figure 2). Moreover, the completeness of the coverage of linkage group VI-C, relative to its diploid homoeolog (dLG 6) pinpointed the location of the sex-determining region at the distal end of VI-C (Figure 2). The macrosynteny of SSR markers combined with the fact that there was still some recombination between the sex-determining loci indicates that the 'proto-sex chromosome' of *F. virginiana* retains a largely autosomal nature.

This comparative mapping approach was further extended to the nearly exclusively dioecious *F. chiloensis* to test the hypothesis that the sex chromosomes of the two sibling species represent two stages of the step-wise evolution [10, 11], that is that the sex determining regions of *F. virginiana* and *F. chiloensis* are syntenous but that *F. chiloensis* has increased recombination suppression between the sexual function genes. Goldberg and colleagues [27] mapped chromosomes in Homeologous Group VI (hereafter HG VI), and found that the sex-determining region in *F. chiloensis* is on a chromosome in this group (VI-A), but not the same one as in *F. virginiana* (VI-C) (Figure 2). This difference was supported by the finding that the sex-determining regions are at the opposite ends of their respective chromosomes (top of VI-C versus bottom VI-A) and that these regions are most closely linked to different SSR markers. These findings indicate that the two species do not share the same sex-determining region as has previously been assumed [18] and that sex chromosome diversity can exist in very recently diverged taxa (~ 0.20 MYO [13]). Furthermore, these findings beg the questions: Did the sex-determining regions arise *de novo* in separate autosomes in *F. virginiana* and *F. chiloensis*? Or did the sex-determining region arise in one autosome and later move to another chromosome in one of the species as the result of rearrangement?

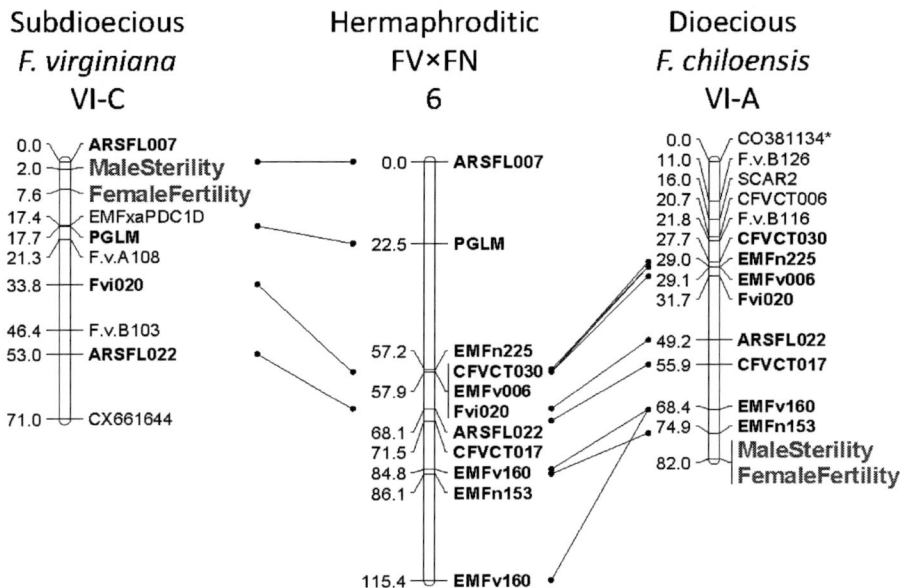

Figure 2. Comparison of the sex-determining chromosomes in subdioecious *F. virginiana* (VI-C, left) and dioecious *F. chiloensis* (VI-A, right) to the homoeologous autosome in a diploid hermaphroditic cross (6, middle). Adapted from [27].

Although rearrangement of the dominant male-sterility allele may be the more parsimonious explanation, the fact that sex chromosomes in the two species are members of the same homeologous group (HG VI) might indicate that certain autosomes are predisposed to becoming sex chromosomes. It has been proposed that some autosomes are 'suited to the task of sex determination' through the existence of genes susceptible to sterility mutations [30], or through the presence of genes that are under sexually antagonistic selection prior to the mutations that cause sterility [31-33]. Indeed, linkage group 6 in diploid *Fragaria* carries many genes involved in reproductive function, including self incompatibility [34], flower and receptacle size [35], anther and ovule number [25], and pollen development [T-L. Ashman, P. Jaisiwal, A. Liston, M. Hanumappa, J. Elser, unpublished data], and thus chromosome 6 could be a highly susceptible autosome.

SECONDARY SEXUAL DIMORPHISM IN FRAGARIA

Sexual dimorphism is the existence of phenotypic differences between the sexes and is thought to arise from selection toward different trait optima via male and female function (e.g., sexually antagonistic selection) [36]. It is often pronounced in animals but also exists in plants to a lesser degree [37, 38]. Here we will refer to this as 'secondary sexual dimorphism' (SSD) to distinguish it from the initial sterility mutations that lead to individuals with different phenotypic genders. Although SSD has not been studied extensively in many *Fragaria* species, significant SSD exists for vegetative and reproductive traits in both *F. virginiana* [6, 37], and *F. chiloensis* [39] (see also Table 3). In this chapter, we present data that suggest there is also a trend for greater SSD in some floral traits in the dioecious compared to the subdioecious or gynodioecious species across four sexually polymorphic *Fragaria* (Figure 3). This pattern is most pronounced for the primary sexual traits of anther number and ovule production per flower. This could indicate that the species with more complete separation of the sexes (e.g., *F. chiloensis*) have been subject to stronger sexually antagonistic selection in the past or that modification of their genetic architecture has lessened a constraint that still limits SSD in species with high frequencies of hermaphrodites (subdioecious *F. virginiana*, gynodioecious *F. vesca* ssp. *bracteata*).

Understanding how a genome shared by both sexes controls sex-specific trait expression is an active area of inquiry. Both sex-determining and autosomal regions of the genome are thought to play roles in the expression of SSD. The sex-determining region can influence expression of SSD if there is linkage between the genes with sex-differential fitness effects to the sex-determining region. This may occur because such linkage can resolve sexual conflict [40-42]. Sex determiners may also interact with autosomal loci [33, 43] to control SSD via sex-limited expression [44-46], or they may act as switches in genetic cascades [47]. QTL mapping has been able to provide initial access to the genetic control of SSD in a few animal systems [47, 48] but our understanding in plant systems still lags far behind.

Table 3. Sexual dimorphism in reproductive and vegetative traits in the *F. virgniana* [10] and *F. chiloensis* [27] mapping populations. Trait means (SE) for the male-sterile and male-fertile F1 progeny as well as *P*-values from *t*-test or Wilcoxon rank sum tests for sexual dimorphism. Sample size for *F. virginiana* is 184 and for *F. chiloensis* 85 progeny.

Trait	*F. virginiana*			*F. chiloensis*		
	Male Sterile	Male Fertile	*P*	Male Sterile	Male Fertile	*P*
Proportion fruit set	0.88 (0.015)	0.09 (0.014)	<0.0001	0.98 (0.005)	0.15 (0.032)	<0.0001
Anther number	20.52 (0.179)	22.89 (0.17)	<0.0001	25.16 (0.46)	25.64 (0.40)	0.44
Proportion seed set	0.71 (0.028)	0.44 (0.026)	<0.0001	0.71 (0.029)	0.12 (0.027)	<0.0001
Ovule number	82.74 (1.394)	80.73 (1.324)	0.30	52.81 (1.601)	54.33 (2.142)	0.57
Flower number	25.14 (1.033)	30.46 (0.982)	<0.0001	13.60 (1.130)	24.05 (2.201)	0.0004
Leaf number	10.86 (0.254)	13.17 (0.241)	<0.0001	17.30 (0.947)	14.48 (0.753)	0.012
Runner number	1.83 (0.076)	1.61 (0.072)	0.032	2.07 (0.191)	2.15 (0.174)	0.61
Plantlet number	4.14 (0.195)	4.31 (0.187)	0.64	2.00 (0.179)	2.23 (0.219)	0.48

In strawberry, the first evidence that there were sex-differences in the genetic architecture came from a quantitative genetics study showing that the genetic variance-covariance matrices (***G***) for females and hermaphrodites differed significantly in three wild populations of *F. virginiana* [6, 37]. Taking the investigation of genetic architecture to the genomic level, Spigler and colleagues [25] recently performed an extensive QTL study in *F. virginiana* which revealed QTL for sexually dimorphic traits both overlapping the sex-determining region and located on autosomes. In addition, this work discovered evidence of sex-limited QTL for proportion seed set. Specifically, two autosomal QTLs reduced seed set in male-fertile plants, but not in male-sterile ones. This study represents a milestone for understanding genetic determination of sexually dimorphic traits in *Fragaria*.

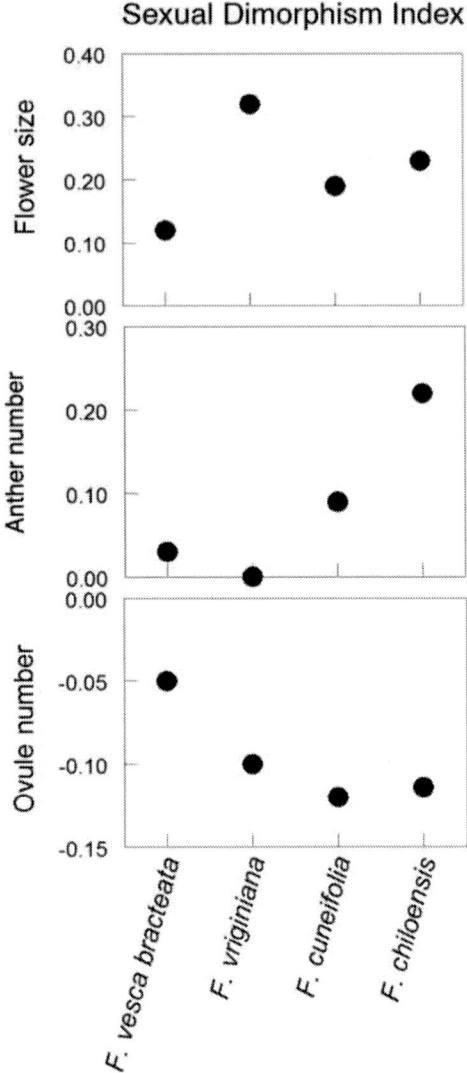

Figure 3. Secondary sexual dimorphism (SSD) in floral traits across four species of *Fragaria*. An index of SSD was calculated for ovule number per flower, anther number per flower and flower size as (mean male fertile - mean male sterile)/mean male fertile. Data are from fresh (floral diameter: *F. chiloensis, F. cuneifolia* [58]; petal size, anther and ovule number: *F. virginiana* [5, 9]) or preserved flowers (flower size, anther and ovule number: *F. vesca* ssp. *bracteata*; anther and ovule number: *F. cuneifolia, F. chiloensis* [T-L. Ashman, M. Koski, K. DeHart, unpublished data]).

Table 4. Summary of QTL analysis for *Fragaria chiloensis* based on results from interval mapping (IM) and multiple-QTL model (MQM) mapping.

Trait	LG[a]	M/P	IM QTL LOD	IM QTL Position	IM PVE	IM A	Markers left/right of QTL peak	Sign. of nearest markers[b]	LOD threshold CL	LOD threshold GW	Cofactor	MQM QTL LOD	MQM QTL position	MQM PVE	MQM A
Proportion fruit set	VI-A	M	42	80.82	94.8	-0.4	EMFn153	*******	1.7	2.4	Male sterility	42	80.82	94.8	-0.43
	VI-C	P					EMFv104/ARSFL22	**							
Fruit number	VI-A	M	6.83	80.82	37.2	-4.3	EMFn153	******	1.6	2.2	Male sterility	6.83	80.82	30.8	-4.27
	VI-B	M	2.12	24.07	14	-2.7	F.V.B126/ARSFL22	****/***	1.6	2.2					
Flower number	VI-A	M	4.7	80.82	28.6	7.1	EMFn153	*****	1.8	2.3	Male sterility	4.7	80.82	28.6	7.12
	VI-C	P	2.13	1.0	12.3	4.6	EMFv104/Fvi20	****/***	1.4	2.2	Fvi20/EMFn153/F.v.C3	2.28	1	10.5	4.33
Leaf number	VI-B	M	1.67	26.91	9.2	1.0	ARSFL22/CFVCT017	****/***	1.5	2.2		1.67	26.91	9.2	-1.63
	VI-C	M	1.4	1.0	7.7	1.5	F.V.B126	**	1.3	2.2		1.4	1	7.7	1.5
Proportion seed set	VI-A	M	23.1	80.82	83	-0.3	EMFn153	*******	1.9	2.5	Male sterility	23.1	80.82	83	-0.32
	VI-B	M	2.06	32.6	14	-0.1	ARSFL22/CFVCT017	***	1.6	2.5					
	VI-B	P					EMFn153	*							
	VI-C	P					ARSFL7	*							
	I(p)	P					EMFn185	*							
Ovule number	VI-A	P	1.81	38.13	11.1	4.0	EMFv104/Fvi20	*******/**	1.7	2.1	Fvi20	1.81	38.132	11.1	4.01
	VI-B	P					F.V.A116	**							
	VI-C	P					F.V.B.103/ARSFL22	**/**							

Table 4. (Continued)

		IM						LOD threshold			MQM			
Trait	LG[a] M/P	QTL LOD	QTL Position	PVE	A	Markers left/ right of QTL peak	Sign. of nearest markers[b]	CL	GW	Cofactor	QTL LOD	QTL position	PVE	A
Anther number	VI-B M	1.14	19.07	7.1	0.8	F.V.B126	**							
Total plantlet number	VI-A M	1.64	21.57	9.2	0.4	CFVCT006/F.V.B11 6	****	1.7	2.3		1.7	21.57	9.6	0.38
	VI-C M					*EMFv104/F.v.B 103*	**							
	VI-A P	2.08	31.79	13.2	-0.4	EMFn225/EMFv104	**/****	1.7	2.2	EMFv104	2.08	31.79	13.2	-0.43
	VI-C P	1.11	5.32	7.2	-0.3	Fvi20/F.V.B.103	**/**	1.4	2.2	EMFv104	2.09	5.32	12.3	-0.40
Total runner number	VI-A M	1.14	21.74	6.5	0.3	F.V.B116	***	1.7	2.3					
	VI-A P	1.71	30.79	10.8	-0.4	EMFn225/EMFv104	**/***	1.8	2.2	F.V.B126/ARSFL22	2.72	51.1	18.6	-1.0
	VI-C P	1.09	5.32	7.3	-0.3	Fvi20/F.V.B.103	**	1.4	2.2	F.V.B126	1.39	6.32	8.7	-0.36

Notes: For each QTL detected, the following information is provided: the linkage group (LG) where the QTL was detected, whether the QTL was found in the maternal (M) or paternal (P) genetic map, LOD value at the QTL peak, the position of QTL peak in centimorgans (cM), the percent variation explained (PVE) by the QTL and its additive effect (A), names of markers flanking the QTL peak, the significance of the associated markers according to single marker (Kruskal-Wallace) analysis, the LOD threshold values at chromosome level (CL) and genome-wide level (GW), the name of marker used as cofactor in MQM analysis along with additional information from MQM results.

[a] QTLs that were significant only in single marker (Kruskal-Wallace) analysis are italicized

[b] $P < 0.1*$, $0.05**$, $0.01***$, $0.005****$, $0.001*****$, $0.0005******$, $0.0001*******$

In this chapter, we present a similar QTL analysis with *F. chiloensis*. We followed the same QTL mapping protocol described in Spigler et al. [25] with the exception that we restricted the QTL search to the mapped chromosomes of HG VI (Table 4). We then compared the results with the QTL on HG VI from Spigler and coworkers [25] to determine whether there were any similarities in QTL location for shared traits, with special regard to those showing SSD. For *F. chiloensis*, interval mapping (IM) identified QTL for all seven traits studied and these were located across three of the four LGs of HG VI (Table 4). The more conservative composite interval mapping (MQM) indicated significant QTL for all of these traits on LG VI-A or VI-C and these are represented in Figure 4 along with the QTLs for the same set of traits on LG VI-A and VI-C in *F. virginiana*.

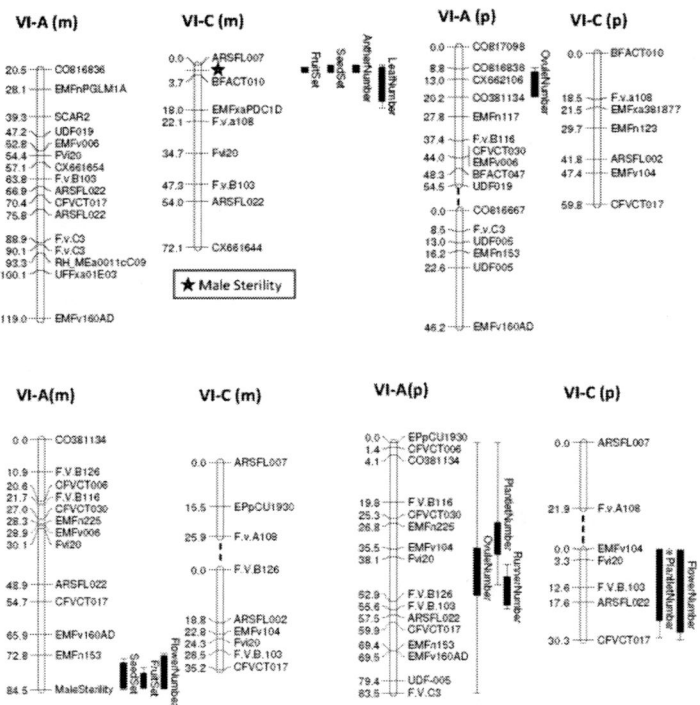

Figure 4. Comparison of QTL for traits in Table 4 that map to sex chromosomes and their homoeologs in *F. virginiana* (top) and *F. chiloensis* (bottom). Chromosomes VI-A and VI-C are presented for maternal (right, denoted by 'm') and paternal (left, denoted by 'p') parents from the *F. virginiana* population [25] and *F. chiloensis* population [27, Table 4]. QTL for traits with an * were only significant at the chromosome level, all others significant at the genome-wide level [25, Table 4].

This comparison revealed several similarities. First, in both mapping populations when female fertility was characterized quantitatively as proportion fruit set it mapped to the single region of the genome that houses the male sterility mutation (Figure 4), confirming the results of qualitative trait mapping [10, 27]. Second, an additional component of female fertility, proportion seed set, also mapped to the respective sex-determining regions in the two species. Third, strongly SSD traits in each species also mapped to their respective sex regions (Table 3, Figure 4). In either case, the QTL overlapping the sex-determining region could represent the physical linkage of genes to that region or that the sex-determining region exerts control of genes for these traits located elsewhere. Teasing apart these mechanisms will require QTL mapping of the sexes separately, and in larger populations than the current sizes of 100-200. Differences between the maps in location of QTL (e.g., anther number and flower number) could suggest differences in genetic control of SSD between the species because, for example, sex-limited expression may evolve from sex-linkage [26, 49]. However, they also could simply reflect differences in the degree of SSD between the mapping populations. A fourth similarity between the species was that both maps show a QTL for ovule number, a sexually monomorphic trait, on LG VI-A (*F. virginiana* also had a QTL on VI-B; [25]). This likely reflects their shared ancestry because there is a QTL for receptacle diameter (an estimator of ovule number) in this same region on LG 6 in the hermaphroditic diploid cross [35]. A final similarity was that there was a QTL detected by IM for seed set at the bottom of VI-A in *F. virginiana* [25], a region homoeologous to the sex-determining region in *F. chiloensis*. While these results, admittedly, are not definitive because of the limited number and size of the mapping populations involved in the comparisons, taken together they do suggest that comparative QTL mapping may be able to help ascertain whether certain autosomes are predisposed to become sex chromosomes or whether there is movement of sex-determining regions between chromosomes, perhaps to other regions where sexually antagonistic genes reside [32].

POLYPLOIDY: KINDLING FOR SEXUAL SYSTEM EVOLUTION?

In several angiosperm lineages, including *Fragaria,* the transition from hermaphroditism to dioecy is associated with speciation via polyploidization [17,

50, 51] suggesting that ecological or genetic aspects of genome doubling/merger facilitates the evolution of separate sexes and possibly sex chromosomes. Miller and Venable [17] proposed that the loss of self incompatibility as a result of genome doubling followed by increased inbreeding depression could facilitate the evolution of separate sexes. It is also possible that other physiological or morphological changes associated with polyploidy such as greater size or drought tolerance [52, 53] could lead to invasion of novel habitats [54], especially those that are conducive to the evolution of separate sexes, e.g., dry or nutrient poor habitats [reviewed in 55]. If genome doubling/merger also leads to an increase in the incidence of sterility mutations [56] or to heritable epigenetic changes due to gene methylation [57] that have a selective advantage under new ecological conditions [11, 55] then there may be an increased chance of gynodioecy evolving from hermaphroditism. Further restructuring of the genome after doubling could also increase the probability of novel gene linkages, such as those between male and female function, thus increasing the chance that sex chromosomes evolve.

These scenarios could apply to *Fragaria* because both genome doubling and expansion into new habitats are characteristics of several of the polyploid species, especially the two octoploid strawberries that have been studied in detail (*F. virginiana* and *F. chiloensis*). Both are believed to be alloallopolyploids (AAA'A'BBB'B'), with likely contributors of the A genomes as ancestors of *F. vesca* s. l. and/or *F. mandschurica* [13, 15] and of the B genome(s) as *F. iinumae* [15]. It is postulated that the two species may only have diverged after migration across the Bering Strait to NW North America from Asia, where *F. chiloensis* spread southward and *F. virginiana* spread eastward and southward, adapting to dry (sand dunes, rocky cliff faces) and moister (meadows, disturbed areas) environments, respectively [12, 58]. If the two species share a male-sterility mutation it is possible that it was the result of a genome merger. Alternatively it is possible that the male-sterility mutation came from a common diploid ancestor related to the subspecies *F. vesca* ssp. *bracteata*, the sole diploid known to have male sterility [18]. In either case, the second sterility mutation, female sterility, which does not to exist in *F. vesca* ssp. *bracteata* [18] but is required to produce a sex chromosome, might have been facilitated by polyploidy. This might be especially likely if chromosome 6 houses genes subject to sterility mutations, because four copies of it would then lead to increased opportunities for the second mutation to occur. Tracing the evolutionary history of the sex-determining region and connecting events to the ecological patterns of migration and divergence will be an exciting endeavor that may be able to shed light on the role of poylpoidy in the evolution of sexual polymorphism in *Fragaria*.

CONCLUSION

The diversity of sexual systems and ploidy levels, presence of female heterogametey and young age make *Fragaria* an extraordinary model genus for understanding the evolution of sex chromosomes. Recent genetic mapping has provided access to sex chromosome identification in two species and fueled ideas for how polyploidy may contribute to sex chromosome evolution. Continued comparative genetic mapping and identification of the male-sterile allele and/or male-sterile-linked DNA sequences will aid in addressing the question of whether male sterility arose once in *Fragaria*, or whether it arose independently multiple times, not only within in the 'octoploid clade' (clade A), but also the clade that houses the sexually polymorphic Asian tetraploids (clade B).

Species of *Fragaria* also provide the opportunity to study the process of recombination suppression in detail. *F. virginiana,* in particular, is a useful system for empirically linking selective and genetic mechanisms that contribute to recombination suppression and thus connect the ecological context for the evolution of dioecy with sex chromosome evolution. If the cessation of recombination between the sex-determining loci is confirmed for *F. chiloensis* then there is opportunity to study its genomic consequences, and especially whether it has lead to heteromorphic sex chromosomes. Nearly a century ago, Kihara [59] reported sex chromosome heteromorphism in *F. moschata* (a.k.a. *F. elatior*), but this observation has not been confirmed nor expanded to other species. Molecular tools currently under development will enable the exploration of sex chromosome homology and morphology across *Fragaria*, and thus provide another perspective to this ecologically-, genetically- and phylogenetically-rich system for understanding sex chromosome evolution.

ACKNOWLEDGMENTS

We thank K. DeHart, D. Jackson, M. Koski, J. Li, B. McTeague, E. York and many past members of the Ashman laboratory for assistance, A. Liston, J. van Ooijen, K. Lewers, D. Sargent for discussion, advice and/or sharing data. This work was supported by funds from the University of Pittsburgh and NSF (DEB 0449488 and 1020523) to TLA.

REFERENCES

[1]   Ming R, Wang J, Moore PH, Paterson AH. Sex chromosomes in flowering plants. *American Journal of Botany*. 2007; 94:141-50.

[2]   Staudt G. The species of *Fragaria*, their taxonomic and geographical distribution. *Acta Horticulturae*. 1989; 265:23-33.

[3]   Stahler MM, Ascher PD, Luby JJ, Roelfs AR. Sexual composition of populations of *Fragaria virginiana* (Rosaceae) collected from Minnesota and western Wisconsin. *Canadian Journal of Botany*. 1995; 73:1457-63.

[4]   Valleau WD. The inheritance of flower types and fertility in the strawberry. *American Journal of Botany*. 1923; 10:259-74.

[5]   Ashman T-L. Determinants of sex allocation in a gynodioecious wild strawberry: implications for the evolution of dioecy and sexual dimorphism. *Journal of Evolutionary Biology*. 1999; 12:648-61.

[6]   Ashman T-L. Constraints on the evolution of dioecy and sexual dimorphism: Field estimates of quantitative genetic parameters for reproductive traits in three populations of gynodioecious *Fragaria virginiana*. *Evolution*. 2003; 57:2012-25.

[7]   Bishop E, Spigler RB, Ashman T-L. Sex-allocation plasticity in hermaphrodites of sexually dimorphic Fragaria virginiana (Rosaceae). *Botany*. 2010; 88:231-40.

[8]   Hancock JF, Bringhurst RS. Hermaphroditism in predominately dioecious populations of *Fragaria chiloensis* (L.) Duchn. *Bulletin of the Torrey Botanical Club*. 1979; 106:229-31.

[9]   Ashman T-L, Hitchens MS. Dissecting the causes of variation in intra-inflorescence allocation in a sexually polymorphic species, *Fragaria virginiana* (Rosaceae). *American Journal of Botany*. 2000; 87:197-204.

[10]   Spigler RB, Lewers KS, Main DS, Ashman T-L. Genetic mapping of sex determination in a wild strawberry, *Fragaria virginiana*, reveals earliest form of sex chromosome. *Heredity*. 2008; 101:507-17.

[11]   Charlesworth B, Charlesworth D. A model for the evolution of dioecy and gynodioecy. *The American Naturalist*. 1978; 112:975-97.

[12]   Potter D, Luby J, Harrison R. Phylogenetic relationships among species of Fragaria (Rosaceae) inferred from non-coding nuclear and chloroplast DNA sequences. *Systematic Botany*. 2000; 25:337-48.

[13]   Njuguna W. Development and use of molecular tools in *Fragaria*. Corvallis: Oregon State University; 2010.

[14]   Njuguna W, Bassil N, Cronn R, Liston A. Phylogenetic analysis of *Fragaria* (Rosaceae) based on nearly complete chloroplast genomes. *Botany*   2009; http://2009.botanyconference.org/engine/search/index.php?func=detail&aid =740.

[15]   Rousseau-Gueutin M, Gaston A, Aïnouche A, Aïnouche ML, Olbricht K, Staudt G, et al. Tracking the evolutionary history of polyploidy in *Fragaria* L.(strawberry): New insights from phylogenetic analyses of low-copy nuclear genes. *Molecular Phylogenetics and Evolution*. 2009; 51:515-30.

[16]   Staudt G. Strawberry biogeography, genetics and systematics. *Acta Horticulturae*. 2009; 842:71-83.

[17]   Miller JS, Venable DL. Polyploidy and the evolution of gender dimorphism in plants. *Science*. 2000; 289:2335-8.

[18]   Ahmadi H, Bringhurst RS. Genetics of sex expression in *Fragaria* species. *American Journal of Botany*. 1989; 78:504-14.

[19]   Staudt G. The genetics and evolution of heteroecy in the genus *Fragaria*. I. Investigations on *Fragaria orientalis*. *Z Pflanzenzuchtung*. 1967; 58:245-77.

[20]   Anthony RD, Hedrick UP. Inheritance of sex in strawberries. *Technical Bulletin of the New York Agricultural Experiment Station*. 1917; 63:3-10.

[21]   Kaul M. Male sterility in higher plants. Springer Verlag: Berlin. 1987;1005:116-17.

[22]   Vyskot B, Hobza R. Gender in plants: sex chromosomes are emerging from the fog. *Trends in Genetics*. 2004; 20:432-8.

[23]   Charlesworth D, Mank JE. The birds and the bees and the flowers and the trees: Lessons from genetic mapping of sex determination in plants and animals. *Genetics*. 2010; 186:9-31.

[24]   Staudt G. The genetics and evolution of heteroecy in the genus *Fragaria*. III Investigations of hexaploid and octoploid species. *Z Pflanzenzuchtg*. 1968; 59:83-102.

[25]   Spigler RB, Lewers K, Ashman T-L. Genetic architecture of sexual dimorphism in a subdioecious plant with a proto-sex chromosome. *Evolution*. 2011; 65:114-1126.

[26]   Charlesworth D, Charlesworth B, Marais G. Steps in the evolution of heteromorphic sex chromosomes. *Heredity*. 2005; 95:118-28.

[27]   Goldberg MT, Spigler RB, Ashman T-L. Comparative genetic mapping points to different sex chromosomes in sibling species of wild strawberry (*Fragaria*). *Genetics*. 2010. 186: 1425-1433.

[28]   Spigler RB, Lewers K, Johnson A, Ashman T-L. Comparative mapping reveals autosomal origin of sex chromosome in octoploid *Fragaria virginiana*. *Journal of Heredity* 2010; 101:S107-17.

[29]   Sargent D, Cipriani G, Vilanova S, et al. The development of a bin mapping population and the selective mapping of 103 markers in the diploid *Fragaria* reference map. *Genome*. 2008; 51:120-7.

[30]   Graves J, Peichel C. Are homologies in vertebrate sex determination due to shared ancestry or to limited options? *Genome Biology*. 2010; 11:205.

[31]   Rice WR. Sex chromosomes and the evolution of sexual dimorphism. *Evolution*. 1984; 38:735-42.

[32]   Van Doorn G, Kirkpatrick M. Turnover of sex chromosomes induced by sexual conflict. *Nature*. 2007; 449:909-12.

[33]   Mank JE. Sex chromosomes and the evolution of sexual dimorphism: lessons from the genome. *The American Naturalist*. 2009; 173:141-50.

[34]   Boskovic R, Sargent D, Tobutt K. Genetic evidence that two independent S-loci control RNase-based self-incompatibility in diploid strawberry. *Journal of Experimental Botany*. 2009.

[35]   Sargent D. A gnetic investigation of diploid *Fragaria*: The University of Reading; 2005.

[36]   Lande R. Sexual dimorphism, sexual selection and adaptation in polygenic characters. *Evolution*. 1980; 34:292-305.

[37]   Ashman T-L. The limits on sexual dimorphism in vegetative traits in a gynodioecious plant. *The American Naturalist*. 2005; 166:S5-16.

[38]   Delph LF. Flower size dimorphism in plants with unisexual flowers. In: Lloyd DG, Barrett SCH, editors. Floral Biology: studies on floral evolution in animal pollinated plants. New York: Chapman & Hall; 1996. p. 217-37.

[39]   Hancock JF, Bringhurst RS. Sexual dimorphism in the strawberry *Fragaria chiloensis*. *Evolution*. 1980; 34:762-8.

[40]   Fisher RA. The evolution of dominance. *Biological Reviews*. 1931; 6:345-68.

[41]   Bull JJ. Evolution of sex determining mechanisms: The Benjamin/Cummings Publishing Company, Inc.; 1983.

[42]   Rice WR. Sexually antagonistic genes: experimental evidence. *Science*. 1992; 256:1436-9.

[43]   Fry J. The genomic location of sexually antagonistic variation: some cautionary comments. *Evolution*. 2010; 64:1510-6.

[44]   Long A, Mullaney S, Reid L, Fry J, Langley C, Mackay T. High resolution mapping of genetic factors affecting abdominal bristle number in *Drosophila melanogaster*. *Genetics*. 1995; 139:1273.

[45]   Perry G, Ferguson M, Sakamoto T, Danzmann R. Sex-linked quantitative trait loci for thermotolerance and length in the rainbow trout. *Journal of Heredity*. 2005; 96:97.

[46]   Delph LF, Arntz A, Scotti-Saintagne C, Scotti I. The genomic architecture of sexual dimorphism in the dioecious plant *Silene latifolia*. *Evolution*. 2010.

[47]   Kopp A, Graze R, Xu S, Carroll S, Nuzhdin S. Quantitative trait loci responsible for variation in sexually dimorphic traits in *Drosophila melanogaster*. *Genetics*. 2003; 163:771.

[48]   Yang X, Schadt E, Wang S, Wang H, Arnold A, Ingram-Drake L, et al. Tissue-specific expression and regulation of sexually dimorphic genes in mice. *Genome Research*. 2006; 16:995.

[49]   Rice WR. Sexually antagonistic male adaptation triggered by experimental arrest of female evolution. *Nature*. 1996; 381:232-4.

[50]   Obbard DJ, Harris SA, Buggs RJA, Pannell JR. Hybridization, polyploidy, and the evolution of sexual systems in *Mercurialis* (Euphorbiaceae). *Evolution*. 2006; 60:1801-15.

[51]   Volz SM, Renner SS. Hybridization, polyploidy, and evolutionary transitions between monoecy and dioecy in *Bryonia* (Cucurbitaceae). *American Journal of Botany*. 2008; 95:1297-306.

[52]   Hodgson J, Sharafi M, Jalili A, Díaz S, Montserrat-Martí G, Palmer C, et al. Stomatal vs. genome size in angiosperms: the somatic tail wagging the genomic dog? *Annals of Botany*. 2010; 105:573.

[53]   Maherali H, Walden A, Husband B. Genome duplication and the evolution of physiological responses to water stress. *New Phytologist*. 2009;184:721-31.

[54]   Ainouche M, Fortune P, Salmon A, Parisod C, Grandbastien M, Fukunaga K, et al. Hybridization, polyploidy and invasion: lessons from Spartina (Poaceae*)*. *Biological invasions*. 2009; 11:1159-73.

[55]   Ashman T-L. The evolution of separate sexes: A focus on the ecological context. In: Harder LD, Barrett SCH, editors. The Ecology and Evolution of Flowers. Oxford, UK: Oxford University Press; 2006. p. 204-22.

[56]   Rieseberg LH, Willis JH. Plant speciation. *Science*. 2007; 317:910-4.

[57]   Johnson L, Tricker P. Epigenomic plasticity within populations: its evolutionary significance and potential. *Heredity*. 2010; 105:113-21.

[58]   Staudt G. Systematics and geographic distribution of the American strawberry species. Taxonomic studies in the genus *Fragaria* (Rosaceae: Potentilleae). Doyle JA, editor. Berkeley: University of California Press; 1999.

[59]   Kihara H. Karyologische studien an *Fragaria* mit besonderer beruksichtigung der geschlechtschromosomen. *Cytologia*. 1930;1:345-57.

[60]   Li J, Koski M, Ashman T-L. Functional analysis of gynodioecy in *Fragaria vesca* ssp. *braceata*. Submitted..

[61]   Hummer KE, Nathewet P, Yanagi T. Decaploidy in *Fragaria iturupensis* (Rosaceae). *American Journal of Botany*. 2009; 96:713-6.

In: New Insights on Plant Sex Chromosomes   ISBN: 978-1-61470-236-8
Editor: Rafael Navajas-Pérez   ©2012 Nova Science Publishers, Inc.

Chapter VI

The Genus *Rumex*: A Plant Model to Study Sex-Chromosome Evolution

Rafael Navajas-Pérez[*]

Departamento de Genética, Facultad de Ciencias, Universidad de Granada,
Campus de Fuentenueva, Spain

ABSTRACT

The origin and evolution of sexual dimorphism and the sex-determining mechanisms are major topics in Evolutionary Biology on which many studies have focused in recent decades. Among flowering plants, the origin of dioecy has resulted from quite recent events, occurring independently in about 7% of the genera. However, only a moderate number of dioecious plant species exhibit chromosome-mediated sex-determination systems. The genus *Rumex* (*Polygonaceae*), with monoecious, gynodioecious, hermaphroditic, and polygamous representatives, and with dioecious species bearing sex chromosomes in different evolutionary stages (homomorphic, XX/XY, XX/XY_1Y_2), has considerably contributed to shed light on this topic.

* Corresponding author: Departamento de Genética, Facultad de Ciencias, Universidad de Granada, Campus de Fuentenueva s/n, 18071. Granada, SPAIN, email: rnavajas@ugr.es

Although still patchy, current knowledge on sex-chromosome evolution has greatly benefited from analyses on species of this group, for which the most significant findings are reviewed in this chapter.

INTRODUCTION

The genus *Rumex* has been generally related to human activity since ancient times and properties conferred by chemical compounds –mainly vitamin C and oxalic acid and oxalates [1]- have benefits both for culinary and ethnopharmacological use [2, 3, 4]. One of the first references of *Rumex* can be found in Exodus 12:8: "*And they shall eat the flesh in that night, roast with fire, and unleavened bread; and with bitter herbs they shall eat it*". Also, Pliny mentioned that the army of Julius Caesar would have been cured of scurvy by using the "*Erba britannica*", identified later as *Rumex aquaticus* L. However, raw plants are toxic due to the high content in oxalates, and in fact some *Rumex* species can be used as antifertility or abortive products [5]. In agriculture, some *Rumex* species have been used as feed crops, and continue to have applications as a dietary source of potential bioactive compounds [6], and recently have been proposed as potential choices for Hg phytoremediation of contaminated soils [7].

In science, ever since Kiara and Ono [8] first described the complex system of sex chromosomes in *Rumex acetosa* L., studies on sex determination have nearly monopolized the research on this group of species. Many scientific studies have focused on *Rumex* because of its biological and evolutionary significance in sexual dimorphism (comprising dioecious, gynodioecious, polygamous and hermaphroditic species; [9]). Thus, Löve [10], Löve and Kapoor [11], Smith [12, 13], Degraeve [14, 15, 16, 17] and Wilby & Parker [18] have contributed significantly to general knowledge of cytogenetic features of the group. Works by Navajas-Pérez et al. [19] and Cuñado et al. [20] are outstanding regarding the chromosomal evolution of the group, while those by Ono [21], Löve, [22], Smith, [23], Ainsworth, [24] and Cuñado et al. [20], are essential for understanding the sex-determination mechanisms. Also, Kurita & Kuroki, [25], Shibata et al. [26, 27, 28] and Navajas-Pérez et al. [29, 30] have greatly contributed to the molecular characterization of Y-chromosome heterochromatin.

This chapter highlights the most important findings regarding sex-chromosome origin and evolution.

MOLECULAR PHYLOGENY AND IMPLICATIONS ON THE EVOLUTION OF DIOECY

Around 200 European, American, and Asian species constitute the genus *Rumex*. Meissner [31] and Willkomm [32] classified the group into four different sections: *Rumex* L. and *Platypodium* Willk., both including hermaphroditic or monoecious annual or perennial herbs, and *Acetosa* Miller and *Acetosella* Fourr., with dioecious, gynodioecious, hermaphroditic or polygamous herbs and shrubs. This classification agrees with the most recent phylogenies based on sistematics proposed by Rechinger [9] and López González [33], who considered the four groups at the subgeneric level, and by Löve and Kapoor [11] or Degraeve [14, 15, 16, 17], who considered four genera, renaming *Platypodium* Willk. as *Bucephalophora* Pau.

In this chapter we will follow classifications by Rechinger [9] and López González [33]. Thus, according to these authors, the subgenus *Rumex* L. would include the vast majority of species (~75%). These species constitute a highly homogeneous group with a basic chromosome number of x=10. Except for the monoecious *Rumex giganteus* W.T. Aiton, *Rumex skottbergii* Degener & I. Degener and *Rumex albescens* Hillebr., endemic to Hawaii [34], all other representatives of the genus *Rumex* are hermaphroditic. Some ubiquitous species such as *Rumex obtusifolius* L. would have evolved also towards monoecy in Hawaiian populations, so that it is thought that an insular effect might apply in these cases [34, 35, 36].

The subgenus *Platypodium* (Willk.) Rech. Fil. is monospecific and constituted exclusively by the hermaphroditic, diploid, x=8, *Rumex bucephalophorus* L. [16].

The subgenus *Acetosella* (Meissner) Rech. Fil. includes two species, *Rumex acetosella* L. and *Rumex graminifolius* Rudolph ex Lamb., and several subspecies all dioecious with a basic chromosome number x=7. Populations of *Acetosella* s.l. varies from diploid to octoploid level, including tetraploid, pentaploid or hexaploid intermediates [33, 37]. The presence of an XX/XY sex-chromosome system has been observed for all representatives of this group [38]. It has also been demonstrated that the sex-determining mechanism is mediated by the presence of an active Y chromosome [10].

The subgenus *Acetosa* includes species with a basic chromosome number ranging from x=10 to x=4. The species with a basic chromosome number of x=10 or 9 are mainly hermaphroditic, polygamous or gynodioecious. Except for *Rumex sagittatus,* x=9, dioecious without differentiated sex chromosomes [17] and *Rumex suffruticosus*, x=8, with a heteromorphic XX/XY sex-chromosome system [20], the majority of dioecious representatives of this subgenus are x=7 and are represented by *Rumex acetosa* and its close relatives. These species form a homogeneous group characterized by similar morphological and karyological features, including an XX/XY_1Y_2 sex-chromosome system plus a sex-determination mechanism based on the X/A balance [17, 18, 22, 23, 24]. Notably, in the section *Americanae* the dioecious *Rumex paucifolius* has an XX/XY system [39] and *Rumex hastatulus* has two chromosomal races, one x=5, XX/XY (Texas race -TXR) and the other x=4, XX/XY_1Y_2 (North Carolina race, NCR) [23]. The latter race has an X/A-based sex-determination mechanism and the former a Y-based one (see **Figures 1** and **2**).

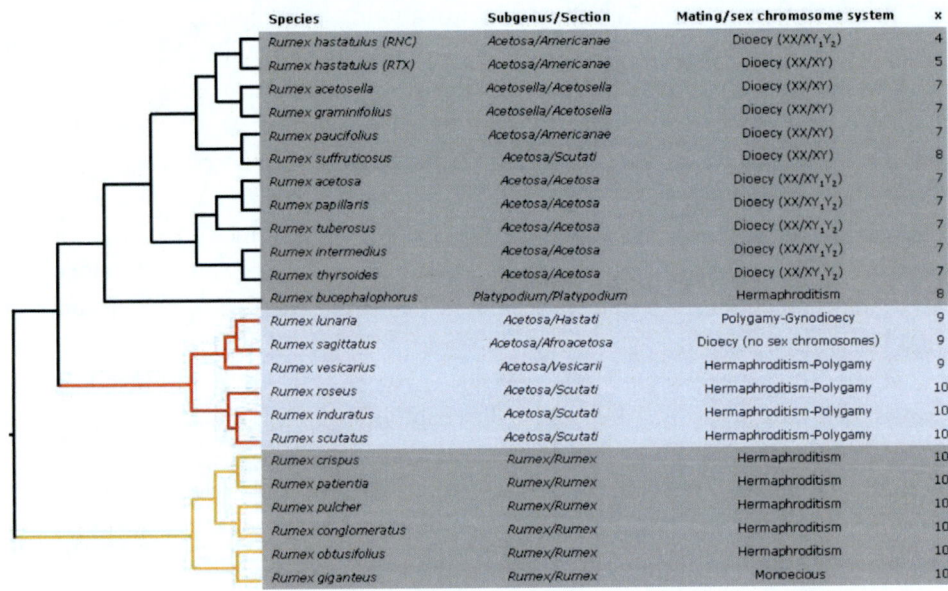

Species	Subgenus/Section	Mating/sex chromosome system	x
Rumex hastatulus (RNC)	*Acetosa/Americanae*	Dioecy (XX/XY_1Y_2)	4
Rumex hastatulus (RTX)	*Acetosa/Americanae*	Dioecy (XX/XY)	5
Rumex acetosella	*Acetosella/Acetosella*	Dioecy (XX/XY)	7
Rumex graminifolius	*Acetosella/Acetosella*	Dioecy (XX/XY)	7
Rumex paucifolius	*Acetosa/Americanae*	Dioecy (XX/XY)	7
Rumex suffruticosus	*Acetosa/Scutati*	Dioecy (XX/XY)	8
Rumex acetosa	*Acetosa/Acetosa*	Dioecy (XX/XY_1Y_2)	7
Rumex papillaris	*Acetosa/Acetosa*	Dioecy (XX/XY_1Y_2)	7
Rumex tuberosus	*Acetosa/Acetosa*	Dioecy (XX/XY_1Y_2)	7
Rumex intermedius	*Acetosa/Acetosa*	Dioecy (XX/XY_1Y_2)	7
Rumex thyrsoides	*Acetosa/Acetosa*	Dioecy (XX/XY_1Y_2)	7
Rumex bucephalophorus	*Platypodium/Platypodium*	Hermaphroditism	8
Rumex lunaria	*Acetosa/Hastati*	Polygamy-Gynodioecy	9
Rumex sagittatus	*Acetosa/Afroacetosa*	Dioecy (no sex chromosomes)	9
Rumex vesicarius	*Acetosa/Vesicarii*	Hermaphroditism-Polygamy	9
Rumex roseus	*Acetosa/Scutati*	Hermaphroditism-Polygamy	10
Rumex induratus	*Acetosa/Scutati*	Hermaphroditism-Polygamy	10
Rumex scutatus	*Acetosa/Scutati*	Hermaphroditism-Polygamy	10
Rumex crispus	*Rumex/Rumex*	Hermaphroditism	10
Rumex patientia	*Rumex/Rumex*	Hermaphroditism	10
Rumex pulcher	*Rumex/Rumex*	Hermaphroditism	10
Rumex conglomeratus	*Rumex/Rumex*	Hermaphroditism	10
Rumex obtusifolius	*Rumex/Rumex*	Hermaphroditism	10
Rumex giganteus	*Rumex/Rumex*	Monoecious	10

Figure 1. List of the most representative species of *Rumex*, indicating phylogenetic relationship according to molecular data (left, based on [19]), affiliation based on morphologic classification, mating/sex-chromosome system and basic chromosome number (x).

Polyploidy is rare in X/A-mediated dioecious *Rumex*, while it has been widely observed in Y-mediated ones [40]. This confirms Muller's theory, who proposed that, contrary to Y-mediated systems an X/A balance sex-determining mechanism will prevent the establishment of dioecious polyploid races in higher animals because of the unbalanced intersexes produced [41].

An assumption of the above classification would imply that dioecy has appeared multiple times through the evolution of *Rumex* directly from hermaphroditic forms. Also, it would support the contention that sex-chromosome evolution would have followed several pathways and that secondary regression to hermaphroditism from dioecious forms would have occurred. Furthermore, this systematic shows no relation between the phylogeny of the group and the evolution of basic chromosome number.

In order to check the veracity of all these assumptions, we performed a phylogenetic analysis in 31 species of the genus by using several molecular markers: from the nuclear genome, the intergenic transcribed spacers (ITS1 and ITS2) between the 18S and the 28S ribosomal genes, and from the chloroplast genome, the intron of trnL gene and the intergenic spacer between this gene and the trnF gene [19]. The analysis considered three criteria: neighbor-joining (NJ), maximum parsimony (MP) and maximum likelihood (ML), implemented by softwares Mega vs2.1 [42] and PAUP* 4.0b10 [43]. In that study, we found no support for the four sub-groups described above. Instead, we found a common origin for all American and Eurasian dioecious *Rumex* species belonging to subgenera *Acetosa* and *Acetosella*, which together with *R. bucephalophorus* of subgenus *Platypodium*, form a well-supported clade (**Figure 1**). This suggests that subgenera *Acetosella* and *Platypodium* might be artificial groups. A second clade includes hermaphroditic, polygamous and gynodioecious species of subgenus *Acetosa*. Also, the dioecious without differentiated sex chromosomes *R. sagittatus,* is included in this latter clade. Finally, a third clade comprises exclusively species belonging to subgenus *Rumex* (**Figure 1**).

In contrast to the current morphological view, this new phylogeny suggests a common origin for all Eurasian and American dioecious species of *Rumex*, with gynodioecy as an intermediate state on the way to dioecy, as the *R. sagittatus* lineage demonstrates. The resulting phylogeny is also consistent with a classification of *Rumex* species according to their basic chromosome number, implying that the evolution of *Rumex* species might have followed a process of chromosomal reduction from x=10 toward x=7 (and finally extending to x=4 in the American lineage) through intermediate stages x=9 and x=8 (**Figure 1**).

The molecular data not only showed that the evolution have followed a main pipeline towards dioecy in the genus *Rumex*, but also support the idea that sex-determining mechanisms based on the balance between the number of X chromosomes and autosomes (X/A balance) has evolved secondarily from male-determining Y mechanisms, as the American lineage of *R. hastatulus* supports. X/Y sex determination is taxonomically more widely distributed than X/A, especially in groups such as fishes and plants, with poorly developed sex-chromosome systems [44], and thus it is assumed that the X/A balance mechanisms evolved secondarily from male-determining Y-chromosome mechanisms [45]. However, there was no direct evidence supporting the proposal that the X/Y system is indeed older until now.

DECIPHERING THE IMPROBABLE

Although the presence of complex sex-chromosome systems has been reported in other animal [46, 47, 48] and plant species [49], it is a rare occurrence and, to our knowledge, its fragmented distribution through a single genus has rarely been proved before. The new molecular phylogeny we proposed, demonstrates that XX/XY_1Y_2 systems have evolved from XX/XY ones through chromosomal rearrangements. Strikingly, our data also support the contention that the multiple sex-chromosomes systems appeared in *Rumex* twice independently; one in the European lineage, *Acetosa section*, and two in the *Americanae* section, *R. hastatulus* NCR (**Figure 2**).

This scenario agrees with the age estimates of *Rumex* sex chromosomes. Using the mean rate of change in plant nuclear DNA of 0.6% per site per million years [50], rDNA ITS mean distance between clades suggests that dioecy appeared in *Rumex* between 15-16 mya, while the divergence time for the *R. acetosella/R. suffruticosus* XX/XY clade leading to the *Acetosa* XX/XY_1Y_2 clade should be 12-13 mya [19]. The split between the two chromosomal races of *R. hastatulus* leading to the XX/XY_1Y_2 NCR system occurred around 600,000 years ago [51]. It is remarkable that in other plant taxa such as papaya [52, 53], *Fragaria* [54, 55] or *Silene* [56, 57, 58, 59, 60] dioecy has also been demonstrated to be an early event in evolution.

Since all dioecious *Rumex* species appear to have a common origin and that simple systems gave rise to complex ones at different points in evolution, the next question we addressed was whether sex chromosomes of both groups have also a common origin or whether they evolved from different autosomal sources. For that task, we thoroughly analysed the molecular structure of Y chromosomes, focusing on repetitive satellite-DNA sequences.

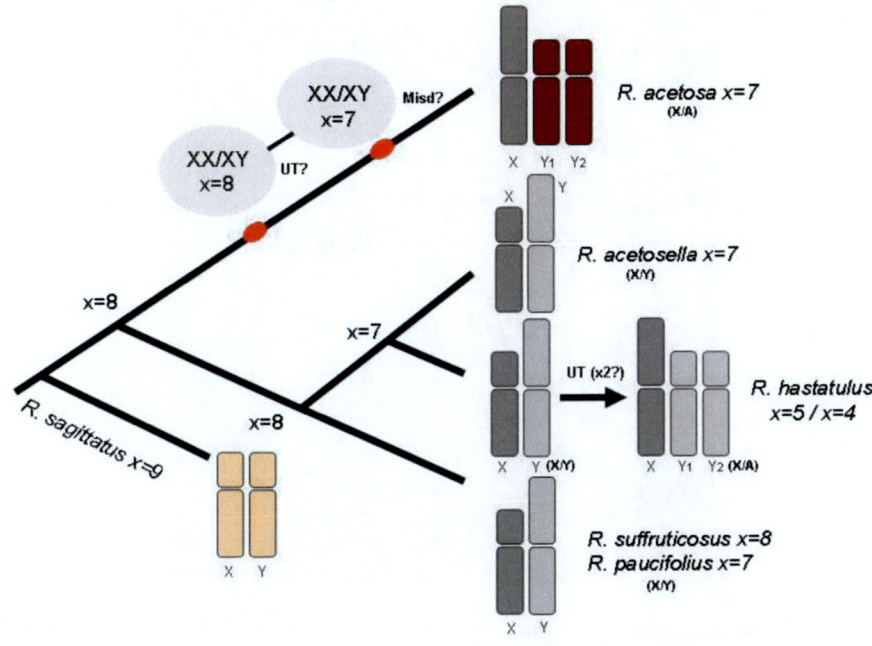

Figure 2. Scheme of sex-chromosome evolution in *Rumex*. *Notes:* there are two possible pathways to explain the origin of complex sex-chromosome systems; by *UT*, unequal translocation from an x=8 ancestor or by *Misd*, misdivision from an x=7 ancestor.

REPETITIVE DNA AND ITS IMPLICATION IN SEX-CHROMOSOME EVOLUTION

The most accepted theories on the origin of sex chromosomes predict that the sexual pair arose form a non-differentiated autosome pair that selectively accumulated the corresponding (male and female) sex-determining genes. To avoid the crossing-over and then the disruption of gender determination, the suppression of recombination can be favoured by methylation or intra-chromosomal rearrangements. In a subsequent stage, the suppression of recombination would spread to surrounding areas and affect the molecular structure of the Y chromosome, which finally would degenerate from the accumulation of repetitive sequences, mainly satellite DNA and transposable elements, mediated by the lack of recombination and Muller's ratchet. A last stage is the dosage compensation in the X chromosome (reviewed in [61, 62]).

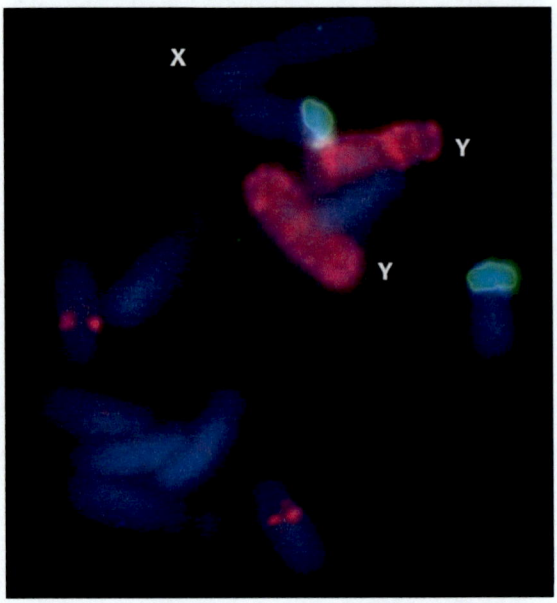

Figure 3. *In situ* hybridization to incomplete root-tip metaphase of *Rumex acetosa* male using as probes RAE180 (red) and RAE730 (green) satellite-DNA sequences.

Repetitive DNA appears to play an important part in the birth and evolution of sex chromosomes. In fact, two satellite-DNA families have been described as being accumulated in the Y chromosomes of several *Rumex* species. On the one hand, RAYSI [26] is a 930-bp repeat satellite exclusively accumulated on both Y chromosomes of XX/XY_1Y_2 species. The same pattern has been found for another two satellite-DNA families, RAYSII and RAYSIII, homologous in sequence with RAYSI [63]. On the other hand, RAE180 [27] is a 180-bp repeat that was demonstrated to be accumulated in a pair of autosomes as well as in the Y chromosomes of *R. acetosa* (**Figure 3**) [20, 27].

We comparatively analysed those families to dissect the molecular structure of sex chromosomes by checking their presence and accumulation in other dioecious species with putative differences in their sex chromosomes. With regard to XX/XY_1Y_2 systems, aside from studying *R. acetosa*, we also studied *R. papillaris, R. intermedius, R. thyrsoides* and *R. tuberosus* (from the section *Acetosa*) and *R. hastatulus* NCR (from the section *Americanae*). With respect to XX/XY systems, we analysed *R. suffruticosus* (from the section *Scutati*) and *R. hastatulus* TXR (from the section *Americanae*). *R. acetosella*, belonging to subgenus *Acetosella* was also analysed (**Figure 1**) [20, 51].

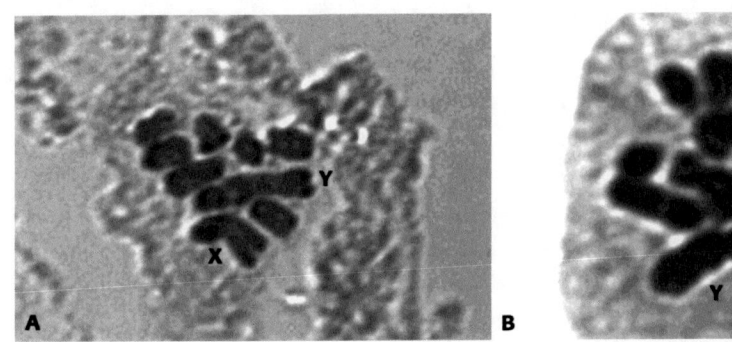
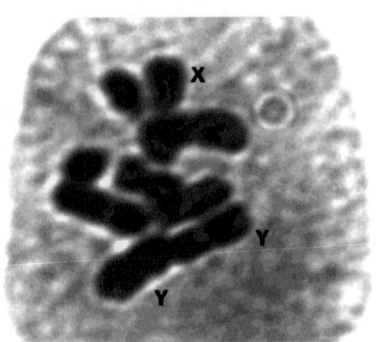

Figure 4. Mitotic chromosomes of *Rumex hastatulus* male; (A) Texas and (B) North Carolina races, showing their sex-chromosome systems.

All species bearing the XX/XY₁Y₂, except for *R. hastatulus* NCR, showed the massive accumulation of RAE180 and RAYSI families in both Y chromosomes that are indeed highly heterochromatic and show massive intra-chromosomal rearrangements (**Figure 3**) [64]. Contrary, the presence of RAYSI in *R. hastatulus* NCR was detected neither by PCR nor by *in situ* hybridization. RAE180 was detected in this species but proved to be very scarcely represented and completely absent from Y chromosomes. In fact, the Y chromosomes of this American endemism remain mostly euchromatic (**Figure 4**) [51]. The XX/XY species showed a similar pattern with the complete absence of RAYSI sequences and the presence of RAE180 in variable amounts but in any case limited to autosomal loci. By DAPI staining *R. hastatulus* TXR and *R. acetosella* Y chromosomes showed no evidence of degeneration, while *R. suffruticosus* exhibited faint signs of molecular differentiation (personal observations).

All these observations would support the proposal that: 1) RAE180 and RAYSI satellite-DNA families have significantly contributed to the origin of Eurasian complex sex-chromosomes systems -XX/XY₁Y₂- and their evolution through the accumulation from autosomal loci to Y chromosomes. RAE180 is present in all American and Eurasian dioecious species and absent from the rest of *Rumex* representatives, while RAYSI is limited exclusively to Eurasian XX/XY₁Y₂ species. Thus, RAE180 would mark the origin of dioecy, while RAYSI would pre-date the origin of complex systems in the Eurasian lineage. 2) Both complex systems -Eurasian and American lineages- evolved independently twice, as we determined above, using molecular markers [19]. This was additionally confirmed by studying another satellite-DNA family -RAE730 [28]. This is absent from sex chromosomes, and is restricted to supernumerary segments of Eurasian XX/XY_1Y_2 species (**Figure 3**), but is not found nor in the American endemic *Rumex* neither in the rest of XX/XY dioecious representatives [29]. Repetitive elements appear to be good makers in this group, being restricted to only closely related species. In this sense, another satellite-DNA family, named RUSI, has been proved to be exclusively represented in *R. scutatus* and *R. induratus*, but not in the rest of species [65](**Figure 1** and 3) On these grounds, although it is the most parsimonious pathway, we cannot affirm that XX/XY Y-chromosomes have the same origin as those from XX/XY_1Y_2 species because we found no sequences shared by both types of sex chromosomes. A fine comparative mapping strategy would be necessary to rule out this aspect.

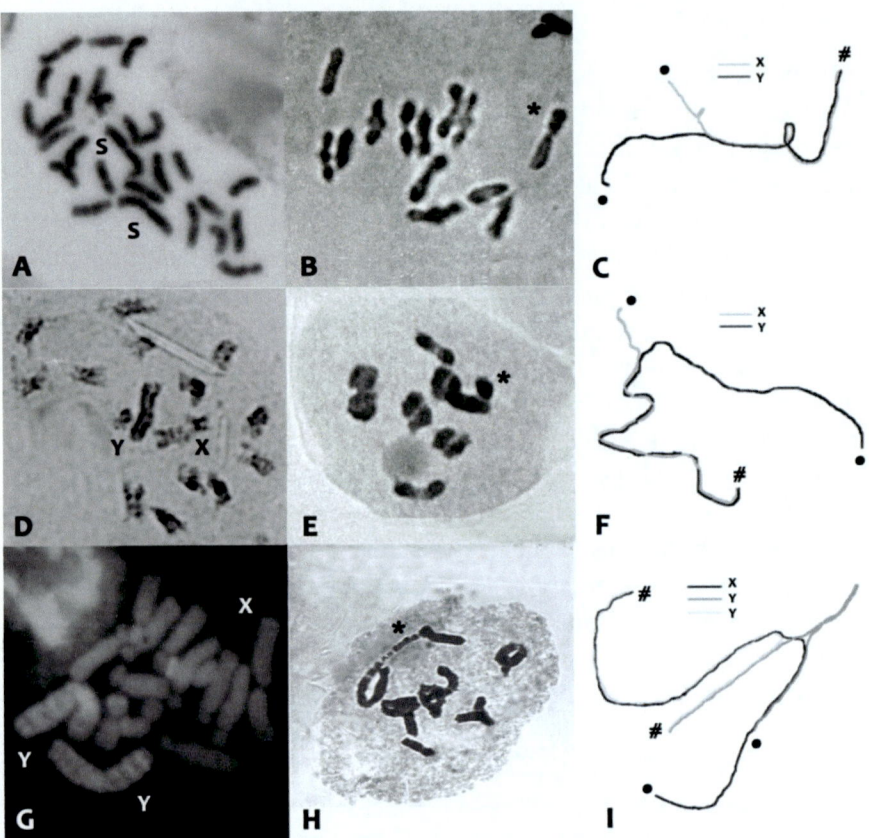

Figure 5. Mitotic and meiotic chromosomes, and sex-chromosome synaptonemal complex ideogram (modified from [20]) of: *Rumex acetosella* (A-C), *Rumex suffruticosus* (D-F) and *Rumex acetosa* (G-I) males. S: indicates sex chromosome, *: sexual bi/trivalent, #: ends of the synaptonemal complex, (black dot) ends of the asynapsed chromosome ends.

TOWARDS SEX-CHROMOSOME DIFFERENTIATION

A combined analysis of both the accumulation of repetitive DNA and the meiotic chromosome features of the most representative species of *Rumex* has demonstrated the existence of sex-chromosomes belonging to three different evolutionary stages in the group [20]. **Figure 5** summarizes the three stages that can be found in *Rumex* sex-chromosomes, namely:

An initial stage would be represented by those species with no signs of degeneration in their Y chromosomes. Thus, *R. acetosella* sex chromosomes consistently formed a monochiasmate heteromorphic bivalent indicating the existence of a pseudoautosomal region between the X and the Y chromosomes; this, however, lacked any contrastable DAPI+ or C+ bands [20] (**Figure 5A-C**). *R. hastatulus* could also be considered within this first stage due to the lack of differentiation in its sex chromosomes. It bears noting that, in this group, *R. hastatulus* TXR (x=5) would have given rise secondarily to *R. hastatulus* NCR (x=4 and XX/XY_1Y_2), most likely by unequal translocation events (**Figure 2 and 4**) [66], without the mediation of molecular degeneration of Y chromosomes. A second stage corresponds to *R. suffruticosus*. This species exhibits some evidence of molecular degeneration in the Y chromosome that stains faintly but positively with DAPI and has a markedly heteromorphic sexual bivalent with an evident area of non-homologous asynapsis [20] (**Figure 5D-F**). A final stage would be represented by *R. acetosa* and their close relatives. This group of species is characterized by a highly differentiated sex-chromosome system XX/XY_1Y_2 in which the Y chromosomes are heteropicnotic and show DAPI+ and C+ bands and are massively degenerated by the accumulation of repetitive sequences (**Figure 3 and 5G-I**) [20, 30, 64, 67]. The cytogenetic differentiation between X and Y chromosomes is also confirmed by the synaptonemal complex analysis. *R. acetosa* males showed six homomorphic bivalents and a sexual trivalent in which each Y chromosome was associated with one of the terminal regions of the X chromosome [20] (**Figure 5G-I**).

REDUCED RATES OF EVOLUTION
IN SEX-CHROMOSOME-LINKED SEQUENCES

Although the role of repetitive sequences has been widely reported in Y-chromosome degeneration (see e.g. [68]), little is known about how this accumulation occurs or about how the absence of recombination affects the subsequent evolutionary fate of the repetitive sequences in the Y chromosome. In order to understand the evolutionary dynamics of sex chromosomes in *Rumex*, we isolated several monomeric units from three satellite-DNA families, RAYSI, RAE730, and RAE180, in two related species, *R. acetosa* and *R. papillaris*, and comparatively analysed mutation and concerted-evolution rates. RAYSI and RAE730 originated from a common ancestral 120-bp unit through replication cycles [29]. This situation allows us to analyse two sequences which had a common origin but which accumulated in significantly different areas of the genome: RAYSI, in non-recombining Y chromosomes; and RAE730, in autosomes. Also, RAE180 sequences, unrelated to either RAYSI or RAE730, represent a hybrid situation, being accumulated both in autosomes and in Y chromosomes.

This analysis revealed that the evolutionary dynamics of sex chromosomes is characterized by slowed-down rates of evolution. Specifically, sequences accumulated in Y chromosomes (RAYSI and Y-linked RAE180 sequences) have evolved at half the rate and undergo lower rates of sequence evolution and homogenization than do satellite DNAs in autosomes, RAE730 and autosomal RAE180 [29]. It bears mentioning that despite of their common origin, RAYSI and RAE730 display this differential pattern. A subsequent analysis of RAE180 in XX/XY species, where these sequences are exclusively autosomic, showed intra-specific sequence homogeneity and inter-specific divergence that is a general pattern of concerted evolution for these repeats. Contrary to the Y-linked loci of RAE180 in XX/XY$_1$Y$_2$ species, where ancestral variability has remained with reduced rates of sequence homogenization and of evolution.

All these data support the contention of the non-recombining nature of Y chromosomes and that the molecular-drive process is significantly affected by mechanisms of non-reciprocal exchange and factors such as location, organization or copy number [69].

CONCLUSIONS

- In contrast to the morphological classification, the phylogeny based on molecular markers (chloroplastidial DNA, rDNA, satellite-DNA, dispersed repetitive DNA) and basic chromosome number evolution, suggests a common origin for all Eurasian and American dioecious species of *Rumex*, with gynodioecy as an intermediate state on the way towards dioecy.

- The origin of dioecy in *Rumex* has been estimated at around 15-16 mya, and the split of regular and complex sex-chromosome systems at 12-13 mya. This is a new piece of evidence that sex chromosomes in plants are evolutionarily young.

- Due to the presence of species with different mating systems and several levels of genetic differentiation between the sex chromosomes, *Rumex* is a model for analysing the evolution of sex chromosomes in plants and their evolutionary history from early stages.

- Satellite DNA (RAE180 family) pre-dates the origin of dioecy and the appearance of XX/XY_1Y_2 systems (RAYSI and RAE730 families). Also these sequences have played an important role in the molecular degeneration of Y chromosomes in complex systems, contributing to the establishment of heteromorphic sex-chromosome systems.

- Despite of their origin, satellite-DNA sequences that had accumulated on non-recombining Y chromosomes underwent lower rates of sequence evolution and homogenization than did those in the autosomes. This implies that mechanisms of non-reciprocal exchange and factors such as location, organization or copy number of repeats significantly affected molecular-drive process.

ACKNOWLEDGMENTS

The author wishes to thank Dr. Roberto de la Herrán and Dr. Carmelo Ruiz Rejón for helpful comments that helped to improve the manuscript. Also I am grateful to Dr. Manuel Ruiz Rejón for kindly providing me the plate of *R. acetosa* meiosis, and to Dr. Juan Luis Santos and Dr. Nieves Cuñado for their help with synaptonemal complex interpretation. RN-P is funded by Plan Propio program of University of Granada, SPAIN.

REFERENCES

[1] Duke, J.A. (1983). Medicinal plants of the Bible. Trado-Medic Books, Owerri, NY.

[2] Rivera, D., Obón, C., Inocencio, C., Heinrich, M., Verde, A., Fajardo, J., Llorach, R. (2005). The ethnobotanical study of local Mediterranean food plants as medicinal resources in Southern Spain. *J. Phys. Pharma. 56*, 97-114.

[3] Rivera, D., Obón, C. (1995). The ethnopharmacology of Madeira and Porto Santo Islands, a review. *J Ethnopharmacology* 46, 73-93.

[4] Peña-Chocarro, L. Zapata Peña, L. (1997). Higos, ciruelas y nueces: aportación de la arqueobotánica al estudio del mundo botánico. *Isturitz, 9*, 679-690.

[5] Solomon, T., Largesse, Z., Mekbeb, A., Eyasu, M., Asfaw, D. (2010). Effect of *Rumex steudelii* methanolic root extract on ovarian folliculogenesis and uterine histology in female albino rats. *Afr Health Sci. 10*, 353-61.

[6] Ferreres, F., Ribeiro, V., Izquierdo, A.G., Rodrigues, M.A., Seabra, R.M., Valentão, P. (2006). *Rumex induratus* leaves: interesting dietary source of potential bioactive compounds. *J Agric Food Chem*, 54, 5782-5789.

[7] Moreno-Jiménez, E., Gamarra, R., Carpena-Ruiz, R.O., Millán, R., Peñalosa, J.M., Esteban, E. (2006). Mercury bioaccumulation and phytotoxicity in two wild plant species of Almadén area. *Chemosphere*, 63, 1969-1973.

[8] Kiara, H. & Ono, T. (1925). The sex chromosomes of *Rumex acetosa. Z. Indukt. Abst. U. Vererb. 39*, 1-7.

[9] Rechinger, K.H. Jr. (1964). The genus *Rumex* L. Pp. 82–89 in T. G. Tutin, V. H. Heywood, N. A. Burges, D. H. Valentine, S. M. Walters, and D. A. Webb, eds. *Flora Europaea*, Vol. 1. Cambridge University Press, Cambridge.

[10] Löve, Á. (1944). Cytogenetic studies on *Rumex* subgenus *acetosella. Hereditas* 30, 1-136.

[11] Löve, Á., Kapoor, B. (1967). A chromosome atlas of the collective genus *Rumex. Cytologia* 32:320–342.

[12] Smith, B.W. (1964). The evolving karyotype of *Rumex hastatulus. Evolution* 18:93–104.

[13] Smith, B.W. (1968). Cytogeography and cytotaxonomic relationships of *Rumex paucifolius. Am. J. Bot. 55,* 673-683.

[14] Degraeve, N. (1980). Contribution à l'étude cytotaxonomique des *Rumex*. III Le genre *Acetosella* Fourr. *Genetica, 54,* 29-34.

[15] Degraeve N. (1975a). Contribution à *l'étude* cytotaxonomique des *Rumex* . I. Le genre *Rumex* L. *sensu stricto. Caryologia, 28,* 187-201.

[16] Degraeve N. (1975b). Contribution à *l'étude* cytotaxonomique des *Rumex* . II. Le genre *Bucephalophora* Pau. *Caryologia, 28,* 203-206.

[17] Degraeve, N. (1976). Contribution à l'étude cytotaxonomique des *Rumex*. IV Le genre *Acetosa* Mill. *La Cellule, 71*, 231-240.

[18] Wilby, A.S., Parker, J.S. (1988). Recurrent patterns of chromosome variation in a species group. *Heredity, 61,* 55–62.

[19] Navajas-Pérez, R., de la Herrán, R., López-González, G., Jamilena, M., Lozano, R., Ruiz-Rejón, C., Ruiz-Rejón, M., Garrido-Ramos, M.A. (2005a). The evolution of reproductive systems and sex-determining mechanisms within *Rumex* (polygonaceae) inferred from nuclear and chloroplastidial sequence data. *Mol BiolEvol., 22,* 1929-39.

[20] Cuñado, N., Navajas-Pérez, R., de la Herrán, R., Ruiz Rejón, C., Ruiz Rejón, M., Santos, J.L., and Garrido-Ramos, M.A. (2007). The evolution of sex chromosomes in the genus *Rumex* (Polygonaceae): identification of a new species with heteromorphic sex chromosomes. *Chromosome Res, 15,* 825-832.

[21] Ono, T. (1935). Chromosomen und Sexualität von *Rumex acetosa*. *Sci Rep Tohoku Imp Univ IV, 10,* 41–210.

[22] Löve, Á. (1957). Sex determination in *Rumex*. *Proc Genet SocCan, 2,* 31-36.

[23] Smith, B.W. (1969). Evolution of sex-determining mechanisms in *Rumex*. *Chromosomes Today, 2,* 172-182.

[24] Ainsworth, C.C., Lu, J., Winfield, M., Parker, J.S. (1999). Sex determination by X: autosome dosage: *Rumex acetosa* (sorrel). In: Ainsworth CC, ed. *Sex Determination in Plants*. Oxford: BIOS Scientific Publishers, pp. 124-136.

[25] Kurita, M. & Kuroki, Y. (1970). Y-chromosome and heterochromatin in *Rumex acetosa*. *Jpn. J. Genet. 45,* 255-260.

[26] Shibata, F., Hizume, M., Kurori, Y. (1999). Chromosome painting of Y chromosomes and isolation of a Y chromosome-specific repetitive sequence in the dioecious plant *Rumex acetosa*. *Chromosoma, 108,* 266-270.

[27] Shibata, F., Hizume, M., Kurori, Y. (2000a). Differentiation and the polymorphic nature of the Y chromosomes revealed by repetitive sequences in the dioecious plant, *Rumex acetosa*. *Chromosome Res., 8,* 229-236.

[28] Shibata, F., Hizume, M., Kurori, Y. (2000b). Molecular cytogenetic analysis of supernumerary heterochromatic segments in *Rumex acetosa*. *Genome, 43,* 391-397.

[29] Navajas-Pérez, R., de la Herrán, R., Jamilena, M., Lozano, R., Ruiz Rejón, C., Ruiz Rejón, M., Garrido-Ramos, M.A. (2005b). Reduced rates of sequence evolution of Y-linked satellite DNA in *Rumex* (Polygonaceae). *J. Mol. Evol., 60,* 391–399.

[30] Navajas-Pérez, R., Schwarzacher, T., de la Herrán, R., Ruiz Rejón, C., Ruiz Rejón, M., Garrido-Ramos, M.A. (2006). The origin and evolution of the variability in a Y-specific satellite-DNA of *Rumex acetosa* and its relatives. *Gene, 368*, 61-71.

[31] Meissner, K.F. (1856). *Rumex*. In: Candolle, A.P.D. (Ed.), Prodromus. Sumptibus Victoris Masson, Parisiis, pp. 41-74.

[32] Willkomm, H.M. (1862). *Polygoneae*. In: Willkomm, H.M., Lange, J. (Eds.), Prodromus Florae Hispanicae. Sumtibus E. Schweizerbart (E. Koch), Stuttgartiae, pp. 279-292.

[33] López González, G. 1990. Género *Rumex* L. Pp. 595–634 in S. Castroviejo, M. Laínz, G. López González, P. Montserrat, F. Muñoz Garmendia, J. Paiva, L. Villar, eds. *Flora Iberica*, Vol. II. CSIC, Real Jardín Botánico de Madrid, Madrid, Spain.

[34] Wagner, W.L., Shannon, R., Herbst, D.R. (1997). Contributions to the flora of Hawaii. VI. *Bishop Mus. Occas. Pap., 48*, 51-65.

[35] Mosyakin, S.L. & Wagner, W.L. (1998). Notes to two alien taxa of *Rumex* L. (*Polygonaceae*) naturalized in the Hawaiian Islands. *Bishop Museum Occasional Papers, 55*, 39-44.

[36] Wagner, W.L., Herbst, D.R., Sohmer. S.H. (1999). *Manual of the flowering plants of Hawaii,* Volume 2. Revised edition. Bishop Museum Special Publication 97. University of Hawaii and Bishop Museum Press, Honolulu, Hawaii.

[37] Singh, R.B. (1968). A dioecious polyploid in *Rumex acetosella*. *J. Hered. 59,* 168-170.

[38] Singh, R.B. (1964*). Cytogenetic studies in Rumex acetosella and its hybrids with Rumex hastatulus with special consideration of sex determination.* Doctoral Thesis, University of North Carolina.

[39] Löve, Á.& Sarkar, N. (1956). Cytotaxonomy and sex determination of *Rumex paucifolius*. *Can. J. Bot. 34*, 261-268.

[40] Σμιτη, B.Ω. (1955). Sex chromosomes and natural polyploidy in dioecious *Rumex*. *J Hered, 46,* 226-232

[41] Muller, H. J. (1925). Why polyploidy is rarer in animals than in plants? *Am. Nat. 59,* 346-353.

[42] Kumar, S., K. Tamura, I. B. Jacobsen, and M. Nei. (2001*). MEGA 2: molecular evolutionary genetics analysis software.* Arizona State University, Tempe.

[43] Swofford, D.L. 1998. PAUP*: phylogenetic analysis using parsimony (*and other methods). Version 4. Sinauer Associates, Sunderland, Mass.

[44] Charlesworth, B. (1996). The evolution of chromosomal sex determination and dosage compensation. *Curr Biol*, *6*, 149-62.

[45] Westergaard, M. (1958).The mechanism of sex determination in dioecious flowering plants. *Adv. Genet.*, *9*, 217-281.

[46] Toder, R., O'Neill, R.J., Wienberg, J., O'Brien, P.C., Voullaire, L., Marshall-Graves, J.A. (1997). Comparative chromosome painting between two marsupials: origins of an XX/XY_1Y_2 sex chromosome system. *Mamm Genome, 8,* 418-22.

[47] Centofante, L., Bertollo, L.A., Moreira-Filho, O. (2006) Cytogenetic characterization and description of an XX/XY_1Y_2 sex chromosome system in catfish *Harttia carvalhoi* (*Siluriformes, Loricariidae*). *Cytogenet Genome Res.* 112:320-4.

[48] de Oliveira, R.R., Feldberg, E., dos Anjos, M.B., Zuanon, J. (2008). Occurrence of multiple sexual chromosomes (XX/XY_1Y_2 and $Z_1Z_2Z_2Z_2/Z_1Z_2W_1W_2$) in catfishes of the genus *Ancistrus* (*Siluriformes: Loricariidae*) from the Amazon basin. *Genetica*, 134, 243-9.

[49] Grabowska-Joachimiak, A., Mosiolek, M., Lech, A., Góralski, G. (2011). C-Banding/DAPI and *in situ* hybridization reflect karyotype structure and sex chromosome differentiation in *Humulus japonicus* Siebold & Zucc. *Cytogenet Genome Res., 132,* 203-11.

[50] Gaut, B.S. (1998). Molecular clocks and nucleotide substitution rates in higher plants. *Evol.Biol, 30,* 93-120.

[51] Quesada del Bosque, E., Navajas-Pérez, R., Panero, J.L., Fernández-González, A., Garrido-Ramos, M.A. (2011). A satellite-DNA evolutionary analysis in the North American endemic dioecious plant *Rumex hastatulus* (*Polygonaceae*). *Genome, in press.*

[52] Yu, Q., Hou, S., Feltus, F.A., Jones, M.R., Murray, J., Veatch, O., Lemke, C., Saw, J.H., Moore, R.C., Thimmapuram, J., Liu, L., Moore, P.H., Alam, M., Jiang, J., Paterson, A.H., Ming, R. (2008a). Low X/Y divergence in four pairs of papaya sex-liked genes. *Plant J.*, *53,* 124–132

[53] Yu, Q., Navajas-Perez, R., Tong, E., Robertson, J., Moore, P.H., et al. (2008b). Recent Origin of Dioecious and Gynodioecious Y Chromosomes in Papaya. *Tropical Plant Biology*, *1*, 49-57.

[54] Ashman, T.L., Spigler, R.B., Goldberg, M.T., Govindarajulu, R. (2011) *Fragaria*: a polyploid lineage for understanding sex chromosome evolution.

In: Navajas-Pérez, R. (ed), *New Insights on Sex Plant Chromosomes*, Nova Publishers, Hauppauge NY.

[55] Goldberg, M.T., Spigler, R.B., Ashman, T.L. (2010). Comparative genetic mapping points to different sex chromosomes in sibling species of wild strawberry (*Fragaria*). *Genetics, 186,* 1425-33.

[56] Filatov, D. (2011). How much do we know about evolution of sex chromosomes in plants? In: Navajas-Pérez, R. (ed), *New Insights on Sex Plant Chromosomes,* Nova Publishers, Hauppauge NY.

[57] Nicolas, M., Marais, G., Hykelova, V., Janousek, B., Laporte, V., Vyskot, B., Mouchiroud, D., Negrutiu, I., Charlesworth, D., Moneger, F. (2005). A gradual process of recombination restriction in the evolutionary history of the sex chromosomes in dioecious plants. *PLoS Biol., 3,* 47-56.

[58] Guttman, D.S., Charlesworth D. (1998). An X-linked gene with a degenerate Y-linked homologue in a dioecious plant. *Nature, 393,* 1009–1014.

[59] Desfeux, C., Lejeune B. (1996). Systematics of Euromediterranean *Silene* (*Caryophyllaceae*): evidence from a phylogenetic analysis using ITS sequences. *C. R. Acad. Sci. III 319,* 351–358.

[60] Desfeux, C., Maurice S., Henry J.P., Lejeune B., Gouyon P.H.. (1996). Evolution of reproductive systems in the genus *Silene*. *Proc. R Soc. Lond. B 263,* 409–414.

[61] Ruiz Rejón, M. (2004).*Sex chromosomes in plants. In: Encyclopedia of Plant and Crop Sciences* (Vol 6): Dekker Agropedia (6 vols), Marcel Dekker Inc., New York, pp. 1148-1151.

[62] Charlesworth, D. (2002). Plant sex determination and sex chromosomes. *Heredity* 88:94–101.

[63] Mariotti, B., Manzano, S., Kejnovský, E., Vyskot, B., and Jamilena, M. (2009). Accumulation of Y-specific satellite DNAs during the evolution of *Rumex acetosa* sex chromosomes. *Mol Genet Genomics, 281,* 249-59.

[64] Navajas-Pérez, R., Schwarzacher, T., Ruiz Rejón, M., Garrido-Ramos, M.A. (2009a). Molecular cytogenetic characterization of *Rumex papillaris*, a dioecious plant with an XX/XY_1Y_2 sex chromosome system. *Genetica, 135,* 87-93.

[65] Navajas-Pérez, R., Schwarzacher, T., Ruiz Rejón, M., Garrido-Ramos, M.A. (2009b). Characterization of RUSI, a telomere-associated satellite-DNA, in the genus *Rumex* (*Polygonaceae*). *Cytogenetic and Genome Research, 124,* 81-89.

[66] Smith, B.W. (1963). The mechanism of sex determination in *Rumex hastatulus. Genetics, 48*, 1265-1288.

[67] Ruiz Rejón, C., Jamilena, M., Garrido-Ramos, M.A., Parker, J.S., Ruiz Rejón, M. (1994). Cytogenetic and molecular analysis of the multiple sex chromosome system of *Rumex acetosa. Heredity, 72,* 209-215.

[68] Kazama, Y., Matsunaga, S. (2011). The role of repetitive sequences in the evolution of plant sex chromosomes. In: Navajas-Pérez, R. (ed), *New Insights on Sex Plant Chromosomes*, Nova Publishers, Hauppauge NY.

[69] Navajas-Pérez, R., Quesada del Bosque, M.E., Garrido-Ramos, M.A. (2009c). Effect of location, organization, and repeat-copy number in satellite-DNA evolution. *Mol. Genet. Genomics, 282,* 395-406.

Index

T

U

V

W

X

Y